Advances in Cotton Research

Edited by Mahmood-ur-Rahman Ansari

Published in London, United Kingdom

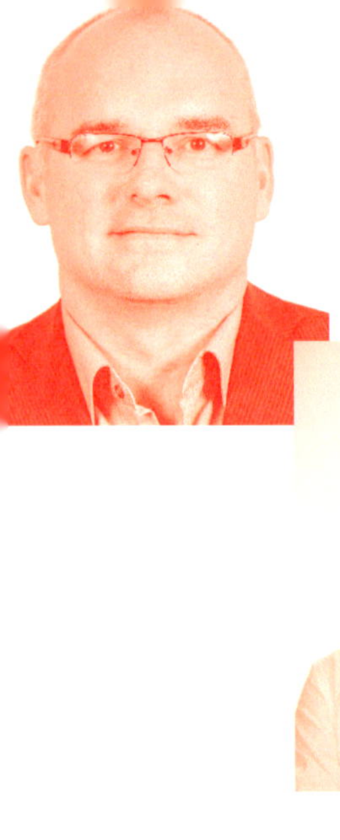

IntechOpen

Supporting open minds since 2005

Advances in Cotton Research
http://dx.doi.org/10.5772/intechopen.77828
Edited by Mahmood-ur-Rahman Ansari

Contributors
Muhammad Rafiq Shahid, Muhammad Shakeel, Muhammad Farooq, Saghir Ahmad, Abid Mahmood, Clara Silvestre, Rafat Al Afif, Christoph Pfeifer, Tobias Pröll, Mahmood-Ur-Rahman Ansari, Parwsha Zaib, Muhammad Iqbal, Roshina Shahzadi, Hafiz Muhammad Ahmad, Bilal Rasool, Aamir Hassan, Muhammad Ijaz

First published in London, United Kingdom, 2020 by IntechOpen
IntechOpen is the global imprint of INTECHOPEN LIMITED, registered in England and Wales, registration number: 11086078, 7th floor, 10 Lower Thames Street, London, EC3R 6AF, United Kingdom
Printed in Croatia

British Library Cataloguing-in-Publication Data
A catalogue record for this book is available from the British Library

Additional hard and PDF copies can be obtained from orders@intechopen.com

Advances in Cotton Research
Edited by Mahmood-ur-Rahman Ansari
p. cm.
Print ISBN 978-1-78984-353-8
Online ISBN 978-1-78984-354-5
eBook (PDF) ISBN 978-1-78984-311-8

For EU product safety concerns:
IN TECH d.o.o., Prolaz Marije Krucifikse Kozulić 3, 51000 Rijeka, Croatia, info@intechopen.com or visit our website at intechopen.com.

Meet the editor

Dr Mahmood-ur-Rahman Ansari is an Assistant Professor of molecular biology in the Department of Bioinformatics and Biotechnology, GC University – Faisalabad, Pakistan. He received his BSc (Hons) degree in plant breeding and genetics from University of Agriculture, Faisalabad, Pakistan in 2003. He received his MPhil and PhD in molecular biology from the National Centre of Excellence in Molecular Biology, Lahore, Pakistan in 2006 and 2011 respectively. He has published over 50 papers in international peer-reviewed journals in the field of molecular biology, biotechnology, and bioinformatics. He has also published 8 book chapters and edited 2 books. He is involved in research projects to understand the molecular mechanisms of stress tolerance in plants, especially cotton and sunflower. He has been involved in the genetic modification of cotton and rice, including their greenhouse and field testing as well as biosafety studies. He is a member of various national and international professional societies. He is a founding member and on the Board of Directors of the Pakistan Society for Computational Biology (PSCB) since 2012. He is also a member of the International Cotton Genome Initiative (ICGI) since 2004.

Contents

Preface

Cotton is one of the most important fiber crops worldwide. It is grown in more than 100 countries, accounting for about 40% of the world fiber market. More than 350 million people are directly or indirectly associated with the cotton production industry. Cotton has a major share in the global agriculture economy; however, its production and yield are compromised due to various biotic, abiotic, and climatic factors. To meet global need, innovative research has been carried out to enhance cotton production. Starting from conventional breeding strategies to modern technologies, there is a long history of cotton research and development. This book presents the milestones of cotton research, including recent and innovative results. It deals with cotton biotechnology, cotton genetics and genomics, genetic modification of cotton, and other topics.

After an introductory chapter providing an overview of recent technological advances in cotton production, Chapter 2 describes the importance of Bt cotton, which is a pest-resistant variety. It also discusses the benefits of Bt cotton adoption and technology. There are lots of insecticides that are used for the management of insect pests in cotton, but these insecticides can be a source of environmental pollution. Bt cotton is resistant against insect pests in an environmentally friendly way. Chapter 3 highlights abiotic stress as a major limitation for crop production. Abiotic stresses include drought, salinity, temperature, and water logging. Chapter 4 describes the use of cotton stalks as a potential raw material for the production of bioenergy. It reviews potential energy generation from cotton stalks through combustion, hydrothermal carbonization, pyrolysis, fermentation, and anaerobic digestion technologies. Chapter 5 discusses the two most important factors influencing cotton spinning: moisture and temperature. The results show that low temperature and high humidity are good for cotton yarn processing.

This book is designed for research students, scientists, academicians, the general public, and other stakeholders. I would like to thank IntechOpen for giving me the opportunity to edit this book. I am also thankful to Ms. Sara Debeuc, Author Service Manager, for her valuable help throughout the editing process. I must thank my research associate, Ms. Parwsha Zaib, for her assistance in the preparation of the introductory chapter and handling the manuscripts. I am especially thankful to all the authors for their valuable contributions to this book.

Mahmood-ur-Rahman Ansari, PhD
Department of Bioinformatics and Biotechnology,
GC University,
Faisalabad, Pakistan

Chapter 1

Introductory Chapter: Recent Trends in "Cotton Research"

Parwasha Zaib, Muhammad Iqbal, Roshina Shahzadi, Hafiz Muhammad Ahmad, Bilal Rasool, Shahbaz Ali Khan and Mahmood-ur-Rahman Ansari

1. Introduction

Cotton is one of the most important fiber crops in the world. It is cultivated in more than 100 countries contributing about 40% of the global market. About 350 million people are directly or indirectly linked with cotton production and industry. Cotton has major share in global agriculture economy; however, its production and yield are compromised due to various biotic, abiotic, and climatic factors. To meet the global need, innovative research has been carried out to enhance the cotton production. Starting from the conventional breeding strategies to the modern technologies, there is a long history of cotton research. In this chapter, the milestones of cotton research have been reviewed with recent and innovative results. The chapter briefly deals with cotton biotechnology, cotton cytogenetics, cotton genetics and genomics, genetic modification of cotton, genome editing, genome sequencing, etc. It contains the latest information in the field of cotton research.

Cotton is major economic crop in the world. However, its production is restricted by various biotic and abiotic factors. Its growth and efficiency are rig orously influenced by the stresses such as high temperature, salinity, less water availability, diseases, insects, etc. The chapter discusses the recent trends in cotton research to combat these biotic and abiotic stresses in crop plants.

2. Abiotic stress tolerance in cotton

Drought is a major abiotic stress which limits plant growth and fiber quality. Yield is badly affected under less water conditions. It is widely spread in many countries and is expected to increase further [1] due to sudden climate changes. Plant growth is decreased due to drought stress as photosynthetic rate and nutrient availability in soil are reduced. Similarly, salinity causes physiological dryness. High temperature can also cause stress [2], which is cause of reduced yield. A lot of research efforts have been done with a statement that the better understanding of these mechanisms would result in crop production under stress conditions. Desiccation stress results in the changing expression pattern of genes in plants, usually arbitrated with abscisic acid (ABA) [3] and physiology of plant is badly affected by the drought stress [4].

When a plant is under stress, the genes are turned on or off. Some proteins are regulatory in nature and regulate signal transduction and regulate expression of stress-responsive genes. Transcription factors also regulate gene expression and are grouped into large families, like WRKY, bZIP, AP2/ERF, Cys2His2, MYC, zinc finger, MYB, and NAC [5]. The DREB subfamily of proteins belongs to the CBF

IntechOpen

(C repeat binding factor) proteins. ERF subfamily also has crucial role in plant gene expression under abiotic stress response and therefore received considerable attention in the recent past [6]. Drought causes significant damage to plants [7–9]. Plants have their self-defense mechanisms which respond to abiotic stresses at molecular, biochemical, and physiological levels [10–14]. Some genes are known to have role in the drought stress response at transcription level [8, 13, 15]. Few drought-tolerant genes have also been studied, which code for regulatory proteins, involved in further regulation of stress-responsive gene expression [14, 15].

Cotton expressing AtNHX1 was grown in the presence of salt, which produced more fiber and biomass. Cotton plants overexpressing the AtNHX1 gene resulted in the movement of salt (Na^+) into the vacuole, which leads to high concentration of salt in the vacuole, which leads to high salt tolerance in cotton. AVP1 gene from Arabidopsis introduced into the cotton resulted in enhanced salt tolerance and more fiber production [16]. CMO gene is a major catalyst in the production of glycine betaine and transgenic cotton expressing AhCMO was more tolerant to high salt concentration then non-transgenic cotton. Thus, it is clear that the salt tolerance in the cotton can be enhanced by the genetic engineering [17]. Cotton plants expressing GhDof1 gene resulted in an increased tolerance to cold and salt stress. GhSOD, GhMYB, and GhP5CS are the stress-responsive genes [18].

3. Biotic stress tolerance in cotton

GhWRKY3 is a gene which is expressed under biotic stress. Cotton seeds were analyzed with three fungi known as *Fusarium oxysporum*, *Rhizoctonia solani*, and *Colletotrichum gossypii* to understand the response of GhWRKY3. The level of GhWRKY3 protein was increased after infection with these three fungi clearly indicated that GhWRKY3 has an important role in fungal pathogenesis [19]. Increased pathogen tolerance can be achieved by the transgenic management of the expression of the GhLac1 gene in cotton (*Gossypium hirsutum*). Elevated tolerance to the fungal pathogen, *Verticillium dahliae*, and cotton aphid (*Aphis gossypii*) resulted in response to the up-regulation of GhLac1.

GhWRKY39 present in the nucleus and few cis-acting elements related to stress response were studied. GhWRKY39 gene was expressed by the fungal and bacterial attack and this gene resulted in the enhanced activity of the SOD, POD, and CAT antioxidants after pathogen attack. Thus, GhWRKY39 was found to be involved in the plant protection against pathogen attack. GhWRKY39 directs the reactive oxygen species (ROS) system, which positively results in regulation of plant protection against pathogens [20].

4. Cell and tissue culture technology

Cotton is included in the list of commercial plants because it can be generated from undifferentiated tissues. Tissue can be taken from plant which can be cultured on an appropriate media. By using tissue culture technology, cotton plants have been screened for disease resistance and embryonic plants produced from the callus young plants produced by suspension cultures from callus. Shoot tips were also used to regenerate whole cotton plants [21]. Cotton has successfully been regenerated from the shoot apex. Problem related to regeneration of plant from callus has been sorted out by using shoot apex as explant [21–23]. Cotton was regenerated by the use of shoot apex which was less expensive and required less labor to regenerate cotton plant.

5. Genetically modified cotton

Lots of efforts have been made to genetically modify cotton for insect tolerance. Bt cotton is the success story in this regard. It is being grown in more than 20 countries for the last more than two decades. Extensive studies on Bt cotton have been done in different countries like India [24], Mexico [25], China [26], and Argentina [27]. Bt cotton verities have good impact on environment and human health. Bt cotton has been produced by the transfer of genes from the bacterium known as *Bacillus thuringiensis* which produce the insecticidal proteins. Bt cotton also resulted in the production of high level of useful insects [28].

Naturally, 15% oleic acids are present in the cotton seeds. So, transgenic cotton was developed to contain high levels of oleic acids. By the subcloning of mutant allele of the Fad2 gene from the phaseolin (seed specific promoter), the level of Fad2 gene in cotton was decreased. Gas chromatography was used to analyze the fatty acids profile of seed lipids from the transgenic cotton varieties. About 21–31% oleic acids were seen in the transgenic lines. Progeny of some transformants showed high levels of oleic acid (47%). So, genetically modified cotton can be developed to produce high level of oleic acid [29].

6. CRISPR/cas9 system in cotton

Genome editing can be used for the functional studies and crop improvements. CRISPR/cas9 system has a sgRNA, which directs the break of double-stranded DNA but all sgRNAs are not equally good. So, in cotton it is necessary to decrease the use of less-efficient sgRNA which can be used in the production of genetically transformed plants without the preferred CRISPR/cas9-induced mutations. The transient expression system was used to improve the functions of sgRNA in cotton. This method was used to check the target sites for the genes known as GhEF1, GhPDS, and GhCLA1 and to analyze the nature of mutation induced by CRISPR/cas9 system. Most frequent mutations observed were deletions. So, it was confirmed that CRISPR/cas9 can generate the mutations in the cotton genes, which are very important for the allotetraploid plant. It was also shown that targeting of gene can be achieved by the expression of many sgRNAs. CRISPR/cas9 was used to generate the deletions in the GhPDS locus. Genetically modified cotton having gene editing mutations in GhCLA1 gene was produced by the CRISPR/cas9 system. Intense albino phenotype was produced by the mutation in the GhCLA1 gene [30].

Cotton is a significant crop for the production of fiber, oil and bio-fuel. Usually, *Agrobacterium tumefaciens*-mediated transformation into cotton takes 8–10 months to generate the T0 plants. Scientists used the transient expression system to validate the CRISPR/cas9 cassettes in cotton. Efficient CRISPR/Cas9 cassettes can be selected to get the better mutagenesis rate by the use of GhU6 promoter instead of Arabidopsis ATU6–29 promoter and GhU6 promoter. When CRISPR/cas9 expressed the sgRNA under the GhU6 promoter, CRISPR/cas9 mutagenesis rate was increased four to six times and expression level of sgRNA was increased from six to seven times, which was a great achievement in the targeted mutagenesis of cotton by the CRISPR/cas9 system [31]. CRISPR/cas9 system has been used to generate multiple sites in *Gossypium hirsutum*. Two genes GhCLA1 and DsRed2 were selected as targets. Plants containing edited DsRed2 gene reverted its character to wild type in T0 generation. Gene editing efficiency was 66–100% [32].

7. Cotton computational analysis

Genomic, genetic, and breeding data are available on the Cottongen (http://www.cottongen.org), which has information related to markers, gene maps, expressed sequence tags, whole genome sequence, and genetic map. Analysis tools like BLAST are also available on the Cottongen [33]. Drought-tolerant cDNA libraries have been made by suppressive subtractive hybridization (SSH) technique and many genes were characterized and expression level of the genes under drought stress was studied [34]. DREB gene was identified in cotton (*Gossypium arboreum*), which was cloned and sequenced. Further, it was characterized in silico to study its interactions with other genes. It was found to interact with MYB, NAC, ABRE, and AREB, which are involved in the drought stress tolerance pathways [35].

Acknowledgements

This work was supported by funds from Higher Education Commission of Pakistan.

Author contributions

PZ, RS, and HMA did the research work and wrote the first draft of manuscript. MR and MI designed and wrote the paper and BR and SAK assisted in writing the paper. All the authors read the manuscript and approved it for publication.

Author details

Parwasha Zaib[1], Muhammad Iqbal[2], Roshina Shahzadi[1], Hafiz Muhammad Ahmad[1], Bilal Rasool[3], Shahbaz Ali Khan[2] and Mahmood-ur-Rahman Ansari[1*]

1 Department of Bioinformatics and Biotechnology, GC University, Faisalabad, Pakistan

2 Department of Environmental Science and Engineering, GC University, Faisalabad, Pakistan

3 Department of Zoology, GC University, Faisalabad, Pakistan

*Address all correspondence to: mahmoodansari@gcuf.edu.pk

References

[1] Burke EJ, Brown SJ, Christidis N, Burke EJ, Brown SJ, Christidis N. Modeling the recent evolution of global drought and projections for the twenty-first century with the Hadley centre climate model. Journal of Hydrometeorology. 2006;**7**(5):1113-1125. DOI: 10.1175/JHM544.1

[2] Chinnusamy V, Schumaker K, Zhu J. Molecular genetic perspectives on cross-talk and specificity in abiotic stress signalling in plants. Journal of Experimental Botany. 2003;**55**(395): 225-236. DOI: 10.1093/jxb/erh005

[3] Chandler PM, Robertson M. Gene expression regulated by abscisic acid and its relation to stress tolerance. Annual Review of Plant Physiology and Plant Molecular Biology. 1994;**45**(1):113-141. DOI: 10.1146/annurev.pp.45.060194.000553

[4] Hanson AD, Witz WD. Metabolic responses of mesophytes to plant water deficits. Annual Review of Plant Physiology. 2003;**33**(3): 163-203. DOI: 10.1146/ANNUREV.PP.33.060182.001115

[5] Umezawa T, Fujita M, Fujita Y, Yamaguchi-Shinozaki K, Shinozaki K. Engineering drought tolerance in plants: Discovering and tailoring genes to unlock the future. Current Opinion in Biotechnology. 2006;**17**(2):113-122. DOI: 10.1016/j.copbio.2006.02.002

[6] Sakuma Y, Liu Q, Dubouzet JG, Abe H, Shinozaki K, Yamaguchi-Shinozaki K. DNA-binding specificity of the ERF/AP2 domain of arabidopsis DREBs, transcription factors involved in dehydration- and cold-inducible gene expression. Biochemical and Biophysical Research Communications. 2002;**290**(3):998-1009. DOI: 10.1006/bbrc.2001.6299

[7] Boyer JS. Plant productivity and environment. Science. 1982;**218**(4571):443-448. DOI: 10.1126/science.218.4571.443

[8] Ingram J, Bartels D. The molecular basis of dehydration tolerance in plants. Annual Review of Plant Physiology and Plant Molecular Biology. 1996;**47**(1):377-403. DOI: 10.1146/annurev.arplant.47.1.377

[9] Jakab G, Ton J, Flors V, Zimmerli L, Métraux J-P, Mauch-Mani B. Enhancing arabidopsis salt and drought stress tolerance by chemical priming for its abscisic acid responses. Plant Physiology. 2005;**139**(1):267-274. DOI: 10.1104/pp.105.065698

[10] Schroeder JI, Kwak JM, Allen GJ. Guard cell abscisic acid signalling and engineering drought hardiness in plants. Nature. 2001;**410**(6826):327-330. DOI: 10.1038/35066500

[11] Shinozaki K, Yamaguchi-Shinozaki K. Gene expression and signal transduction in water-stress response. Plant Physiology. 1997;**115**(2):327-334. DOI: 10.1104/pp.115.2.327

[12] Xiong L, Schumaker KS, Zhu J-K. Cell signaling during cold, drought, and salt stress. The Plant Cell. 2002;**14**(Suppl. 1): S165-S183. DOI: 10.1105

[13] Zhu J-K. Salt and drought stress signal transduction in plants. Annual Review of Plant Biology. 2002;**53**(1):247-273. DOI: 10.1146/annurev.arplant.53.091401.143329

[14] Shinozaki K, Yamaguchi-Shinozaki K. Gene networks involved in drought stress response and tolerance. Journal of Experimental Botany. 2007;**58**(2):221-227. DOI: 10.1093/jxb/erl164

[15] Shinozaki K, Yamaguchi-Shinozaki K, Seki M. Regulatory network of gene

expression in the drought and cold stress responses. Current Opinion in Plant Biology. 2003;**6**(5):410-417. Available from: http://www.ncbi.nlm.nih.gov/pubmed/12972040

[16] Pasapula V, Shen G, Kuppu S, et al. Expression of an arabidopsis vacuolar H+-pyrophosphatase gene (AVP1) in cotton improves drought- and salt tolerance and increases fibre yield in the field conditions. Plant Biotechnology Journal. 2011;**9**(1):88-99. DOI: 10.1111/j.1467-7652.2010.00535.x

[17] Ma X, Dong H, Li W. Genetic improvement of cotton tolerance to salinity stress. African Journal of Agricultural Research. 2011;**6**(33): 6798-6803. DOI: 10.5897/AJARX11.052

[18] Su Y, Liang W, Liu Z, et al. Overexpression of GhDof1 improved salt and cold tolerance and seed oil content in *Gossypium hirsutum*. Journal of Plant Physiology. 2017;**218**:222-234. DOI: 10.1016/j.jplph.2017.07.017

[19] Guo R, Yu F, Gao Z, An H, Cao X, Guo X. GhWRKY3, a novel cotton (*Gossypium hirsutum* L.) WRKY gene, is involved in diverse stress responses. Molecular Biology Reports. 2011;**38**(1):49-58. DOI: 10.1007/s11033-010-0076-4

[20] Shi W, Liu D, Hao L, Wu C, Guo X, Li H. GhWRKY39, a member of the WRKY transcription factor family in cotton, has a positive role in disease resistance and salt stress tolerance. Plant Cell, Tissue and Organ Culture. 2014;**118**(1):17-32. DOI: 10.1007/s11240-014-0458-8

[21] McCabe DE, Martinell BJ. Transformation of elite cotton cultivars via particle bombardment of meristems. Nature Biotechnology. 1993;**11**(5): 596-598. DOI: 10.1038/nbt0593-596

[22] Gould J, Banister S, Hasegawa O, Fahima M, Smith R. Regeneration of *Gossypium hirsutum* and *G. barbadense* from shoot apex tissues for transformation. Plant Cell Reports. 1991;**10**(1):12-16. DOI: 10.1007/BF00233024

[23] Saeed NA, Zafar Y, Malik KA. A simple procedure of *Gossypium meristem* shoot tip culture. Plant Cell, Tissue and Organ Culture. 1997;**51**(3):201-207. DOI: 10.1023/A:1005958812583

[24] Naik G. An analysis of socio-economic impact of Bt technology on indian cotton farmers. In: Centre for Management in Agriculture. Ahmedabad, India: Indian Institute of Management; 2001. p. 19

[25] Traxler G, Godoy-Avilla S, Falck-Zepeda J, Espinoza-Arellano JJ. Transgenic cotton in Mexico: Economic and environmental impacts. In: Paper Presented at the Fifth International Conference, Biotechnology, Science and Modern Agriculture: A New Industry at the Dawn of the Century, Ravello, Italy. 2001. pp. 15-18

[26] Pray C, Rozelle S, Huang J, Wang Q. Plant biotechnology in China. Science. 2002;**295**:674-677

[27] Qaim M, de Janvry A. Bt cotton in Argentina: Analyzing adoption and farmers' willingness to pay. 2002 Annual Meeting July 28-31, Long Beach, CA. 2002. Available from: https://ideas.repec.org/p/ags/aaea02/19710.html [Accessed: 23 July 2019]

[28] Fitt GP, Mares CL, Llewellyn DJ. Field evaluation and potential ecological impact of transgenic cottons *(Gossypium hirsutum)* in Australia. Biocontrol Science and Technology. 1994;**4**(4): 535-548. DOI: 10.1080/09583159409355367

[29] Chapman KD, Austin-Brown S, Sparace SA, et al. Transgenic cotton plants with increased seed oleic acid content. Journal of the American Oil

Chemists' Society. 2001;**78**(9):941-947. DOI: 10.1007/s11746-001-0368-y

[30] Gao W, Long L, Tian X, et al. Genome editing in cotton with the CRISPR/Cas9 system. Frontiers in Plant Science. 2017;**8**:1364. DOI: 10.3389/fpls.2017.01364

[31] Long L, Guo D-D, Gao W, et al. Optimization of CRISPR/Cas9 genome editing in cotton by improved sgRNA expression. Plant Methods. 2018;**14**(1):85. DOI: 10.1186/s13007-018-0353-0

[32] Wang P, Zhang J, Sun L, et al. High efficient multisites genome editing in allotetraploid cotton (*Gossypium hirsutum*) using CRISPR/Cas9 system. Plant Biotechnology Journal. 2018;**16**(1):137-150

[33] Yu J, Jung S, Cheng C-H, et al. CottonGen: A genomics, genetics and breeding database for cotton research. Nucleic Acids Research. 2014;**42**(D1):D1229-D1236. DOI: 10.1093/nar/gkt1064

[34] Zhang L, Li F-G, Liu C-L, Zhang C-J, Zhang X-Y. Construction and analysis of cotton (*Gossypium arboreum* L.) drought-related cDNA library. BMC Research Notes. 2009;**2**:120. DOI: 10.1186/1756-0500-2-120

[35] Shaheen T, Fazal Abbas M, Zaib P, Ullah I, Tahir ul Qamar M, Arif A. Identification, characterization, homology modeling and protein-protein interactions of cotton (*Gossypium arboreum* L.) DREB gene. International Journal of Agriculture and Biology. 2018;**20**(5):1055-1061

Chapter 2

Tritrophic Association between Bt Cotton, Arthropod Pest and Natural Enemies

Muhammad Rafiq Shahid, Muhammad Shakeel, Muhammad Farooq, Saghir Ahmad and Abid Mahmood

Abstract

Benefits and harmful effects of Bt adoption technology are mainly related with cotton production where lot of insecticides are needed for management of arthropod herbivory and possible negative impact of crystalline Bt protein on parasitoids and predators is real. Therefore, current review information was focused that Bt should be selective for natural enemies and information was collected from different sources especially CAB abstracts as well as citations from many review articles and books. Usefulness of integrated pest management was highlighted with updated literature to cover the contents.

Keywords: biosafety, transgenic crop, beneficial fauna, non-target arthropod, parasitoid and non Bt refugia

1. Introduction

Development of transgenic crops is by incorporation of Bt genes that are very useful for the effective control of lepidopterous larvae [1]. Bt crops include maize, cotton, tobacco, eggplant, tomato, potato, canola, cabbage, soybean, cabbage and broccoli [2–5]. Cultivated area of biotech crops have also been increased to (2 billion hectares in 2015) that was 67.7 million during 2003 [6, 7]. Biological activity of Bt against target insect pest succeeds as the most effective pest management strategy that can be integrated into pest management module of an economic crop [8–10]. Bt crops have environmentally advantage over microbial Bt without its economic drawbacks because they use sun light energy to manufacture toxin and they are available to manage herbivorous arthropod pest throughout the season in the plant [11]. They are also useful because their toxin is active only against a narrow range of target pest as compared with conventional broad-spectrum insecticides (**Table 1**).

1.1 Detection of Bt protein

It depends mainly upon the toxin level produced by a plant of specific cultivar. However, Bt expression is fluctuated with respect to plant part and the growing stage of the crop [13, 14]. At flowering stage an abrupt change in Bt protein has been recorded [15]. Physiological changes in Bt concentration also occur when plants are damaged [16]. Secondary metabolites alter the effectiveness of Bt protein;

Toxin	Year	Bollworms	Event
Cry-1Ac	1996	*Helicoverpa*, *Pectinophora*, *Earias* spp.	Bollgard® Delta and pineland/ Monsanto
Cry1Ac + Cry2Ab	2003	*Helicoverpa*, *Pectinophora*, *Earias*, *Spodoptera* spp.	Bollgard-II® Delta and pineland/Monsanto
Cry1Ac + Cry1Ab and or CpTi	1997	*Pectinophora*, *Earias* spp.	Bollgard-II® Delta and pineland/Monsanto
Cry1Ac + Cry1Fa	2005	*Helicoverpa*, *Pectinophora*, *Earias*, *Spodoptera* spp.	Wide Strike® Dow Agro Sciences
Vip3A	2005	*Helicoverpa*, *Pectinophora*, *Earias*, *Spodoptera*, *Agrotis* spp.	Vip Cot® Syngenta

Modified from Tabashnik et al. [12].

Table 1.
Generations of Bt and their potential benefits.

terpenoids and condensed tannins both alter the efficacy of Cry1Ac protein. Hydrolysable tannins combined with Bt toxins increased the mortality of lepidopterous larvae [17, 18]. Environmental stresses also can alter the efficacy of Bt protein in the field.

1.2 Factors affecting efficiency of Bt technology

Target larvae of lepidopterous species uptake different amount of Bt toxins by feeding on Bt-expressing parts of the plant [19]. Obrist et al. [20] also found that thrips, *Frankliniella tenuicornis* acquired Cry1Ab after feeding on Bt plants. Mellet and Schoeman [21] quantified Bt concentration in the body of insects with the help of enzyme linked immunological sorbent assay (ELISA) method. Sucking arthropods extract phloem but uptake very minute quantity of Bt toxin as compared with lepidopterous larvae [19, 22].

1.3 Bio-safety detection of transgenic crops

Pest species is suppressed by naturally occurring biological and environmental agents. Living organisms are found to kill, weaken or reduce the reproductive potential of pest species. Effectiveness of biological control agents is enhanced by "conservation" of biocontrol agent, i.e., to protect and maintain population of naturally existing biocontrol agents. Conservation techniques include reducing or eliminating insecticide applications. If the number of biocontrol agents are less than the pest species population existing in a locality, the biocontrol agents can be reared in laboratory and released to increase their population in that area. This is called as "augmentation of biocontrol agents." Sometimes few number of beneficial insects are needed to be released in several locations to suppress population of local insect pest, i.e., called inoculative release and in other case large numbers are released on a single locality, i.e., called as inundative release. If the population of biocontrol agents exists high in one locality they can also be redistributed to the new locality where there is pest problem and beneficial fauna is low. The biocontrol agents can also be imported from one country to the country where there is pest problem. All these techniques are very useful from pest management point of view and have very useful results as biological control against invertebrate pest species. Effectiveness of biocontrol agents is affected by environmental resistance/climatic adaptability, synchrony with life cycle of insect pest, insecticidal and Bt toxin effects.

Benefits and risks of transgenic crops are common concerns of Bt technology. Potential benefits are reduction in use of broad-spectrum insecticides. It produced an opportunity for conservation of biological control agents. It has less hazardous effects on non-target arthropods than insecticides. Its long term cultivation also had no significant impact on soil health [23]. However, effect of Bt insecticidal proteins on beneficial insects is given in **Table 2**

1.4 Risks of Bt on target and non-target species

However, there are also some putative risks related with development of Bt resistance in target insect pest [24] and direct or indirect effect on non-target arthropod biodiversity [25]. Pollen drift by out crossing in open pollinated crops and release of Bt toxin into soil through root exudation, natural wounding or root cell senescence and its effect on soil health as well as its degradation are the other risks of Bt technology [26].

1.5 Effect of Bt on predators

Little change in pest community has ability to change the community of associated beneficial fauna of natural enemies. It is due to direct feeding on transgenic plant tissue and indirectly due to change in food quality by emission of volatile exudates that can attract or repel the pest/non targeted arthropod fauna and ultimately affect the food-web [27]. Cry1Ac toxin of Bt plant is transferred through non-target herbivores to natural enemies at different trophic levels. Therefore, harmful effects on natural enemy population are appeared [28]. It is due to change in food-web [27]. Predators population is reduced either due to change in tritrophic interaction, i.e., shortage of pray or ill effect of Cry protein on beneficial [29].

1.5.1 Heteropteran predators

Geocoris punctipes and *Orius insidiosus* species are common and important members of the natural enemies of a variety of food crops worldwide. They are omnivorous feeders and their generalized feeders of sucking insect pest and acquire Cry1Ab from the body of insect herbivory [21, 30]. However, uptake of Cry1Ac toxin differed with respect to size and feeding requirement of the predators. It was comparatively high in spined soldier bugs, *Podisus maculiventris* that consumed considerably more prey biomass compared to small predatory heteropterans (*G. punctipes* and *O. insidiosus*) [31, 32]. Insecticidal Bt protein exerted adverse

Crop	Bt protein	Targeted arthropod	Tritrophic effect on Non-targeted arthropod	Detrimental effect	References
Cabbage	Cry1Ac	*Plutella xylostella*	Parasitoid	Not detected	[33]
Cabbage	Cry1Ac	*P. xylostella*	*Chrysoperla carnea* larvae	Detected	[33]
Rice	Cry1Ac, Cry2Aa or Cry1Ca	Brown Planthopper	Pond wolf spider	No	[34]
Cotton	Cry1Ac	*Helicoverpa armigera*	*C. carnea* larvae	Not detected	[31, 32]

Table 2.
Tritrophic impact of Bt on populations of the major beneficial predators.

effect on the biology of beneficial fauna. When heteropteran predators fed on *Spodoptera exigua* collected from Cry1Ac cotton field, effect of Bt on biological activity of predators were observed and longevity of *Orius tristicolor* and *Geocoris punctipes* was reduced as compared to non-transgenic treatments.

1.5.2 Effect of Bt on lady bug

Coccinellid beetles belong to very important and most diverse group of insects that have very useful predatory potential. Birch et al. [35] reported that when lady bugs consumed aphids that had picked Bt for 2–3 weeks, fecundity, egg viability and longevity ladybird longevity was reduced up to 51%. Adverse effects on ladybird reproduction, caused by eating peach-potato aphids from transgenic potatoes, were reversed after switching ladybirds to feeding on pea aphids from non-transgenic bean plants. These results demonstrated that expression of a lectin gene for insect resistance in a transgenic potato line can cause adverse effects to a predatory ladybird via aphids in its food chain. Zhao et al. [36] reported added Bt toxin Cry1Ah and Cry2Ab toxin to artificial diet of *Aphis gossypii* and same aphids were used as feed of coccinellid beetle to investigate the tritrophic effect of Bt on beetle. They reported that development, survival and pupae formation of predatory beetle was not affected by Bt crystalline protein.

1.5.3 Effect of Cry1Ac in spider

Cry1Ac toxin uptake was comparatively high in spined soldier bugs, *Podisus maculiventris* that consumed considerably more prey biomass compared to small predatory heteropterans (*G. punctipes*, *N. roseipennis* and *O. insidiosus*) [31, 32]. According to [34] transgenic Bt rice lines producing Cry1Ac, Cry2Aa or Cry1Ca had no detrimental effects on pond wolf spider (PWS), *Pardosa pseudoannulata*. However, Zhou et al. [37] reported that the activities of three key metabolic enzymes were significantly influenced in spider after feeding on fruit flies containing Cry1Ab.

1.5.4 Effect of BT on Chrysoperla

Larval stage of *C. carnea* has very useful predatory potential and rank an important status from pest management point of view. Mellet and Schoeman [21] reported 55.6% survivorship of *C. carnea* larvae after feeding *Spodoptera littoralis* reared on Bt-free isoline. However, the survivorship of *C. carnea* larvae fed *S. littoralis* reared on Bt was significantly reduced to 17.7%, indicating that there may be an additional negative effect of consuming the intoxicated prey source. Similarly Feeding effect of Bt prolonged development of *C. carnea* and increased larval mortality as reported by Romeis et al. [38]. However, Tian et al. [39] reported that Bt crystalline protein had no direct and indirect effect on *Chrysoperla*.

1.5.5 Effect of BT on rove beetle

Riddick and Barbosa [40] monitored numbers of *Lebia grandis*, a predatory ground beetle that specializes on Colorado potato beetle. The numbers of *L. grandis* were significantly smaller in transgenic and mixed fields. Garcia-Alonso et al., [41] reported no effects of exposure to the toxin Cry1Ab through Bt maize fed-prey on the performance and digestive physiology of the predatory rove beetle *Atheta coriaria*.

1.5.6 Effect of BT on pollinator

Honey bees are important cotton pollinators that visit the flowers of the same plant or randomly the flowers of several other plants and one-third of the crops are insect-pollinated. Non targeted insects are directly and indirectly affected by the toxin produced by the transgenic plants [42]. Likewise, honey bees are exposed to direct exposure of Bt during nectar collection and ingestion of contaminated pollen [43].

1.6 Effect of Bt on parasitoids

Parasitoids population is reduced either due to shortage of pray or ill effect of Cry protein [29]. Cui and Xia [44] also demonstrated that parasitoid population was highly reduced in the Bt cotton plots. Host larvae emerged from Bt protein had prolonged larval duration but with less weight of larvae, pupae and adult similarly emergence of parasitoids and adult longevity were also negatively affected on such diet [45]. Baur and Boethel [46] reported that development of *Cotesia marginiventris* inside the body of *Pseudoplusia includens* larvae fed on Bt cotton had reduced longevity and oviposition. Sometimes it becomes difficult for parasitoids larvae to complete development either due to the premature death of larvae or behavior of prey (Schuler et al., 1999); [47–49]. Baur and Boethel [46] reported that *Cotesia marginiventris* developing inside *Pseudoplusia includens* larvae after feeding on Bt cotton had reduced longevity and females with less number of eggs. Parasitoid larvae also could not complete their development because Bt-susceptible hosts were not able to survive on Bt leaves [48, 49].

1.7 Risks of Bt resistance in herbivorous arthropods

Bt technology has altered the arthropod community by reducing population of targeted lepidopterous larvae. This disturbance is mainly due to unavailability of prey for the predator/parasitoids. Mellet and Schoeman [21] reported highest mortality and delay in development of lepidopterous larvae after feeding on Bt as compared with control non-Bt plants. For clarity of results researchers have suggested that the effect of transgenic crops on arthropod community should be monitored on long term basis by using Simpson waiver index (SWI). This index is used to find out species richness [2].

This wide spread adoption of Bt by growers have also resulted in development of resistance in targeted arthropods to Bt crops and Bt sprays in the field [50]. Now many targeted lepidopteron species have field evolved resistance problem [51, 52]. *Plutella xylostella*, *Spodoptera exigua*, *Helicoverpa armigera* and *Pectinophora gossypiella* are the most common ones with field evolved resistance to Bt [53–56].

Bt-susceptible larvae of *Spodoptera exigua* from commercial non Bt-cotton fields contained around 2.7-fold less toxin in their bodies than resistant larvae from Bt. Similarly, *Tetranychus urticae* contained approximately 10-fold more toxin after feeding on Bt [57] (**Table 3**).

1.8 Use of eco-friendly, integrated pest management techniques

The suppression of pest population by the use of all suitable ways either through prevention, observation and intervention of arthropod pest species. Prevention means to manage pests rather than to eliminate them. In this case initial severity of pest is reduced through crop rotation, change in cropping pattern, plant breeding, changing of planting and harvesting time, use of trap crops as well as clean cultivation. Observation is related with pest scouting, harmful and beneficial insect

USA	Bollgard 1	2000	[58]
India	Bollgard 1	2003	[59]
China	Bollgard 1	1997	[53–56, 60]
Australia	Bollgard 1	2004	[61]
Pakistan	Bollgard 1	2009	[62]

Table 3.
Reports of Cry1Ac resistance in the world.

identification, decision of ETL (economic threshold level) and to determine when and what actions should be taken. Intervention is adaptation of various methods used to reduce the effect of economically damaging pest population. They all are the basic components of IPM (integrated pest management) including use of biological control, physical, cultural and chemical control.

1.9 Cultural control

It is the deliberate manipulation of environment to make it less suitable for the pest by eliminating its food, shelter and create hindrance to multiply insect pest on economic crops. The detail of cultural practices is given below.

1.9.1 Crop rotation

Rotations of crops and clean cultivation are the best cultural management tactics which can be used for the management of insect pest species. Sometimes movement of infested plants from one area serves as source of carryover to a new locality [63]. Preventive action that creates unfavorable condition for the pest will help to overcome such problem. More difficult is to decide cropping pattern year after year. Growing of same crop each year increases chance of pest problem. Crop cultivation serve as food chain of the arthropod herbivory and both arthropod population is interconnected with crop cultivation and planting date of the crops. It has been reported that highest population of arthropod species were present in the continuous cropping fields as compared with fallow fields. Fallow lands were used to break the life cycle of arthropod herbivory due to food shortage and un-availability of host plant [64]. Available crop also affects the feeding behavior of targeted arthropods. Many number of arthropods are monophagous (feed on single crop) or oligophagous (having two or narrow range of host plants). They will be naturally died due to unavailability of food by using crop rotation technique.

1.9.2 Trap cropping

Trap cropping technique practiced before the beginning of modern synthetic insecticides, is making a revival comeback now a day in many countries of the world [65]. The unique feature of trap crop is that it provides more attraction as food source and oviposition site than to the original economic crop [66]. Trap crops have been tested against many arthropod herbivores. In a dual choice test of plants, aphids preferred to lay eggs on brassica rather than wheat indicating that brassica has potential to serve as a trap crop for the management of wheat aphid [45]. Similarly, collard crop served as highly preferred for oviposition of *P. xylostella* than cabbage and it laid 300 times more eggs than on cabbage [67]. Trap crops are therefore recommended as an important tool for the management of arthropod herbivory and to protect the economic crops.

1.9.3 Planting and harvesting time

Change in planting and harvesting time creates discontinuity in food supply for the insect pest species. This technique is called as "phenological asynchrony." Crops are matured before or after the onset of insect pest incidence. So they escape insect pest attack and farmers easily manage insect pests on economic crop. Similarly planting density and plant spacing can affect the pest population and searching behavior of insect pest for food and oviposition site.

1.9.4 Induced resistance

Plant nutrition can influence the feeding, longevity and fecundity of arthropod herbivory. Some macro and micronutrients enhance the resistance mechanism of the plants against insect pest species. They also help the help the plant to compensate damage caused by insect pest. Fertilizer and irrigation application also help the plant to overcome environmental and arthropod feeding damage stress.

1.9.5 Use of plant resistance

The useful technique for the control of insect pests is plant resistance, rather than Bt property resistant plants have many physical and chemical characteristics that are used to overcome insect pest problem. It has been reported that resistant plants/varieties are less infested by insect pest herbivory than susceptible ones. In this way population of insect pest are managed on resistant crops and it is due to the food preference property of the pest [68, 69]. The other biological parameters of the pest are also affected on resistant plants due to presence of primary and secondary metabolites [68]. In this way they provide natural control of the pest without use of pesticides. Resistant plants are safe for natural enemies so also conserve and promote beneficial fauna [70].

1.10 Monitoring of pest population

Forecasting, monitoring and pest scouting are considered as important part to devise an IPM strategy. These techniques are used to determine the population and stage of insect pest infestation as well as the population of biocontrol. These are very useful techniques to decide action plan on the basis of economic threshold level of the pest and existing population of the predators. Pest monitoring is done by using various traps like pheromone, light, colored, sticky, pitfall and suction traps [71]. All of the traps have their unique importance because pheromone traps based on sex attraction are sensitive and species specific, light trap attract general pest and non-target flying insects, colored and sticky traps attract insect pest toward and their specific color, pitfall traps are used to trap soil dwelling insects and suction traps are used to suck minute soft bodied insects [72, 73].

1.10.1 Pheromone traps

However, sex pheromone traps are considered as valuable tools from monitoring point of view [74]. Due to specificity and least hazardous for non-target species pheromone traps are considered the best as compared with chemical control and are acceptable among various alternatives control. The first pheromone trap as a mating disruptant was used by [75] from *Bombyx mori*. Sex pheromone against pink bollworm, *Pectinophora gossypiella* was developed during 1970s. Chemists have identified 30 compounds with properties of pheromones and now they are commercially available for more than 300 targeted arthropod herbivory [76]. They are packaged in slow-release dispensers used as lures in traps for mass trapping of sexually active adults. In this way they create mating disruption. Their efficiency pheromone traps depend upon the ratios of the active component, dose rates and dispensers applied. They are very valuable source of insect pest monitoring because more male moths can be captured in trap even when population is at initial stage [77, 78].

1.11 Refuge crop for pest

A genetic change in population occurs due to the mortality of susceptible arthropod herbivory induced by continuous exposure of Bt toxin. Some out of these organisms survive due to natural tolerance against Bt toxin. Breeding of tolerant insects with each other develop ability to survive and changed to become a whole resistant population. In order to tackle the Bt resistance problem, growing of at least 5% area of non-Bt refugia around Bt crop is recommended. Basically non-Bt refuge allows the targeted arthropods to survive and reproduce. So resistance alleles would be suppressed by susceptible alleles. The lepidopterous insects that express resistance and survived in Bt crop would have chance to mate with susceptible ones grown on non Bt refugia. Susceptible generation produced on non Bt crop will compete for food, shelter and mating with resistant strain of targeted arthropod herbivory. The final result will be reduction in multiplication rate of insect therefore chances for development of susceptible strain are increased after many generations. Seed companies are now using technique of non Bt seed mixing with the Bt crops to suppress problem of Bt resistance. Commercial packets of Bt crop seed also contain premixed non-Bt seeds.

1.12 Refugia for beneficial arthropods

Uncultivated land can sustain a diverse range of beneficial fauna has long been known. Refugia may be located within the cultivated crop and outside the crop. Within crop natural refugia may exist in the form of unsprayed crop area, protected parts of the plant and alternative host plants. Outside the crop field borders, live fences, mixed cropping, intercropping and strip planting have been used to provide refugia for natural fauna. Generalist predators and parasitoids move among these crops depending upon the availability of their host (prey). However, effective use of refugia normally requires crop-pest and natural interaction. Such type of interaction

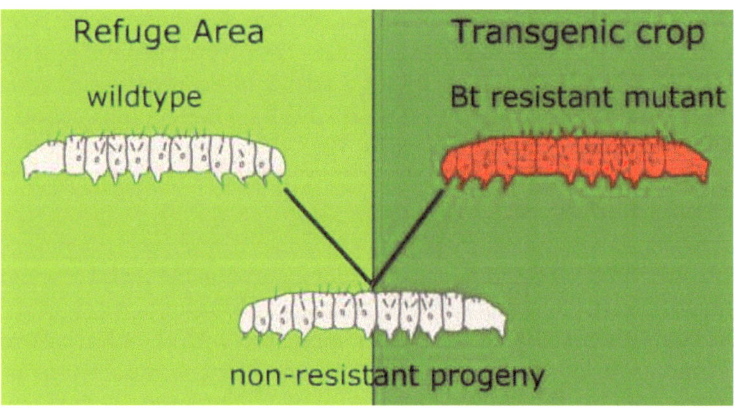

Figure 1.
Importance of non Bt refugia in reducing Bt resistance in targeted larvae.

is called as Tritrophic interaction. Effectiveness of interaction depends upon the time and space because floral composition of refugia, its location and area dimension as well as synchronization of pest and beneficial fauna play an important role from pest management point of view. Planting of *Medicago sativa* strips in cotton has been reported to increase the predatory fauna especially the beetles, *Chrysoperla carnea* that was 5–7 times more abundant on *Medicago* strips than on cotton. This beneficial fauna then migrated to cotton and its population was double on cotton with *Medicago* plots in comparison with cotton without. Refugia have dual function from pest management point of view it has ability to serve as trap crop for pest and conservation of natural enemies. It has also potential to serve as wind speed breaker and overwintering sites for parasitoids. Some natural enemies overwinter within the parts of the plants like wise predatory mite *Typhlodromus* spp., overwinter in calyx activities of apple fruit. The damaged fruits serve as refugia for it and should be left on ground during winter to enhance the population of this type of predatory mite (**Figure 1**).

2. Conclusion

Concerns of bio-safety has been and should be a compulsory component for the long lasting sustainability of transgenic/Bt technology. As mentioned earlier negative impact of Bt on some predators and parasitoids are actual therefore mode of action of Cry1Ac on predators should be investigated by testing of binding and pore formation capacity of toxins to epithelial vesicles of gut membrane, because activity of Cry1Ac toxin is specific for a particular predator. Further food preference difference of predator between Bt and non Bt crop should also be investigated. Difference between availability of semiochemicals in transgenic and conventional crops should also be evaluated because they have tendency to repel the beneficial arthropods. Population of predatory fauna is also badly affected in Bt plots due to reduction of food prey (arthropod herbivory). The key point of the review paper is that Bt crops are more dangerous for the parasitoids than predators. Therefore, we should focus our research on tritrophic impact of Bt crops on parasitoids. Moreover, it was concluded that refugia play an important role from pest management point of view; it can mitigate field evolved resistance in arthropod herbivory on one hand and can sustain the natural parasitoid/predatory fauna on the other hand.

3. Study Questions

1. What is Bt and its significance in pest management? Discuss in the light of modern outlook.

2. Discuss about different generations of Bt.

3. Enlist the factors affecting efficacy of Bt.

4. Elaborate about biosafety detection of transgenic crops for beneficial fauna.

5. Discuss about the Bt resistance problem in target insects.

6. Importance of Integrated Pest Management.

7. Useful of Non Bt Refugia.

Author details

Muhammad Rafiq Shahid[1*], Muhammad Shakeel[3], Muhammad Farooq[1], Saghir Ahmad[1] and Abid Mahmood[2]

1 Cotton Research Institute, Multan, Pakistan

2 Ayyub Agricultural Research Institute, Faisalabad, Pakistan

3 South China Agriculture University, China

*Address all correspondence to: shahid1364@yahoo.com

References

[1] Carozzi N, Koziel M, editors. Advances in Insect Control: The Role of Transgenic Plants. London: Taylor and Francis; 1997

[2] Flint HM, Henneberry TJ, Wilson FD, Holguin E, Parks N, Buehler RE. The effects of transgenic cotton, *Gossypium hirsutum* L., containing *Bacillus thuringiensis* toxin genes for the control of the pink bollworm, *Pectinophora gossypiella* (Saunders) and other arthropods. South West Entomology. 1995;**20**:281-292

[3] Hutchison WD, Bartels DW, Bolin PC. Bt sweet corn, European corn borer and corn earworm: First impressions. In: Proceedings of Midwest Food Processors Association; 9; 1997. pp. 23-26

[4] Jenkins JN, McCarty JC Jr, Buehler RE, Kiser J, Williams C, Wofford T. Resistance of cotton with delta-endotoxin genes from *Bacillus thuringiensis* var. kurstaki on selected lepidopteran insects. Agronomy Journal. 1997;**89**:768-780

[5] Lynch RE, Wiseman BR, Plaisted D, Warnick D. Evaluation of transgenic sweet corn hybrids ex-pressing CryIA (b) toxin for resistance to corn earworm and fall armyworm (*Lepidoptera: Noctuidae*). Journal of Economic Entomology. 1999;**92**:246-252

[6] James C. Global status of commercialised biotech crops. In: International Service for the Acquisition of Agric-Biotech Application. 2015. Available from: http://www.isaaa.org/resources/publications/briefs/49/

[7] James C. Global Status of Commercialized Transgenic Crops: 2003. Ithaca, New York, USA: ISAAA Briefs N° 30. International Service for the Acquisition of Agri-biotech Applications (ISAAA); 2003

[8] Fitt GP, Wilson LJ. Genetic engineering in IPM: Bt cotton. In: Kennedy GG, Sutton T-B, editors. Emerging Technologies for Integrated Pest Management. St. Paul, MN: APS Press; 2000. pp. 108-125

[9] Wilson LJ, Bauer LR, Lally DA. Effect of early season insecticide use on predators and outbreaks of spider mites (*Acari:Tetranychidae*) in cotton. Bulletin of Entomological Research. 1998;**88**:477-488

[10] Wu K. IPM in cotton. In: Jia S, editor. Transgenic Cotton. Beijing, China: Science Press; 2001. pp. 218-224

[11] Hilder VA, Boulter D. Genetic engineering of crop plants for insect resistance—A critical review. Crop Protection. 1999;**18**:177-191

[12] Tabashnik BE, Gould F, Carrière Y. Delayingevolution of insect resistance to transgenic crops by decreasing dominance and heritability. Journal of Evolutionary Biology. DOI: 10.1111/j.1420-9101.2004.00695.x

[13] Daly JC, Olsen KM. The potential for transgenic cotton plants to select for resistance in *Helicoverpa armigera*. In: Proceedings, 9th Australian Cotton Conference; 14-16 August 1996; Broadbeach, Queensland: Australian Cotton Growers Research Association, Wee Waa, Australia; 1996. pp. 295-298

[14] Neal JJ. Brush border membrane and amino acid transport. Archives of Insect Biochemistry and Physiology. 1996;**32**:55-64

[15] Fitt GP, Daly JC, Mares CL, Olsen K. Changing efficacy of transgenic Bt cotton patterns and consequences. In: Zalucki MP, Drew RAI, White GG, editors. Pest Management and Future Consequences. Vol. 1. Brisbane, Australia: University of Queensland Printery; 1998. pp. 189-196

[16] Stout MJ, Workman KV, Duffy SS. Identity, spatial distribution, and variability of induced chemical response in tomato plants. Entomologia Experimentalis et Applicata. 1996;**79**:255-271

[17] Gibson DM, Gallo LG, Krasnoff SB, Ketchum REB. Increased efficacy of *Bacillus thuringiensis* subsp. kurstaki in combination with tannic acid. Journal of Economic Entomology. 1995;**88**:270-277

[18] Morris ON, Converse V, Kanagaratnam P. Chemical additive effects on the efficacy of *Bacillus thuringiensis* Berliner subsp. kurstaki against *Mamestra configurata* (*Lepidoptera*: *Noctuidae*). Journal of Economic Entomology. 1995;**88**:815-824

[19] Head GP, Brown CR, Groth ME, Duan JJ. Cry1Ab protein levels in phytophagous insects feeding on transgenic corn: Implications for secondary exposure risk assessment. Entomologia Experimentalis et Applicata. 2001;**99**:37-45

[20] Obrist LB, Klein H, Dutton A, Bigler F. Ef-fects of Bt maize on *Frankliniella tenuicornis* and exposureto prey-mediated Bt toxin. Entomologia Experimentalis et Applicata. 2005;**115**:409-416

[21] Mellet MA, Schoeman AS. Effect of Bt-cotton on chrysopids, ladybird beetles and their pray: Aphids and whiteflies. Indian Journal of Experimental Biology. 2007;**45**:554-562

[22] Raps A, Kehr J, Gugerli P, Moar WJ, Bigler F, Hilbeck A. Immunological analysis of phloem sap of *Bacillus thuringiensis* corn and of the non-target herbivore *Rhopalosiphum padi* (*Homoptera*: *Aphididae*) for the presence of Cry1Ab. Molecular Ecology. 2001;**10**:525-533

[23] Zhaolei L, Naishun B, Jun C, Xueping C, Manqiu X, Feng W, et al. Effects of long-term cultivation of transgenic Bt rice (Kefeng-6) on soil microbial functioning and C cycling. Scientific Reports. 2017;7(1):4647

[24] Gould F. Sustainability of transgenic insecticidal cultivars: Integrating pest genetics and ecology. Annual Review of Entomology. 1998;**43**:701-726

[25] Conner AJ, Glare TR, Nap JP. The release of genetically modified crops into the environment. Part II. Overview of ecological risk assessment. The Plant Journal. 2003;**33**:19-46

[26] Zwahlen C, Hilbeck A, Nentwig W. Field decomposition of transgenic Bt maize residue and the impact on non-target soil invertebrates. Plant and Soil. 2007;**300**:245-257

[27] Whitehouse MEA, Wilson LJ, Fitt GP. A comparison of arthropod communities in transgenic Bt and conventional cotton in Australia. Environmental Entomology. 2005;**34**:1224-1241

[28] Ives AR, Kareiva P, Perry R. Response of a predator to variation in prey density at three hierarchical scales lady beetles feeding on aphids. Ecology. 1993;**74**(7):1929-1938

[29] Chilcutt CF, Tabashnik BE. Independent and combined effects of *Bacillus thuringiensis* and the parasitoid *Cotesia plutellae* (*Hymenoptera*: *Braconidae*) on susceptible and resistant diamondback moth (*Lepidoptera*: *Plutellidae*). Journal of Economic Entomology. 1997;**90**(2):397-403

[30] Torres JB, Ruberson JR. Canopy- and ground-dwelling predatory arthropods in commercial Bt and non-Bt cotton fields: Patterns and mechanisms. Environmental Entomology. 2005;**34**:1242-1256

[31] Torres JB, Ruberson JR. Interactions of Bt-cotton and the omnivorous

big-eyed bug *Geocoris punctipes* (say), a key predator in cotton fields. Biological Control. 2006;**39**:47-57

[32] Torres JB, Ruberson JR. Spatial and temporal dynamics of oviposition behavior of bollworm and three of its predators in Bt and non-Bt cotton fields. Entomologia Experimentalis et Applicata. 2006;**120**:11-22

[33] Wei M, Russell DW, Mallinckrodt B, Vogel DL. The experiences in close relationship scale (ECR)-short form: Reliability, validity, and factor structure. Journal of Personality Assessment. 2007;**88**:187-204

[34] Niu L, Ma W, Lei C, Jurat-Fuentes JL, Chen L. Herbicide and insect resistant Bt cotton pollen assessment finds no detrimental effects on adult honey bees. Environmental Pollution. 2017;**230**:479-485. DOI: 10.1016/j.envpol.2017.06.094

[35] Birch ANE, Geoghegan IE, Majerus MEN, McNicol JW, Hackett CA, Gatehouse AMR, et al. Tri-trophic interactions involving pest aphids, predatory 2-spot ladybirds and transgenic potatoes expressing snowdrop lectin for aphid resistance. Molecular Breeding. 1999;**5**:75-83

[36] Zhao Y, Zhang S, Luo J-Y, Wang C-Y, Lv L-M, Wang X-P, et al. Bt proteins Cry1Ah and Cry2Ab do not affect cotton aphid *Aphis gossypii* and ladybeetle *Propylea japonica*. Scientific Reports. 2016;**6**(1):1-12

[37] Zhou J, Zhang L, Chang Y, Lu X, Zhu Z, Xu G. Alteration of leaf metabolism in Bt-transgenic rice (*Oryza sativa* L.) and its wild type under insecticide stress. Journal of Proteome Research. 2012;**11**(8):4351-4360. DOI: 10.1021/pr300495x

[38] Romeis J, Dutton A, Bigler F. Bacillus thuringiensis toxin (Cry1Ab) has no direct effect on larvae of the green lacewing *Chrysoperla carnea* (Stephens) (*Neuroptera*: *Chrysopidae*). Journal of Insect Physiology. 2004;**50**:175-183

[39] Tian J-C, Long L-P, Wang X-P, Naranjo SE, Romeis J, Hellmich RL, et al. Using resistant prey demonstrates that Bt plants producing Cry1Ac, Cry2Ab, and Cry1F have no negative effects on *Geocoris punctipes* and *Orius insidiosus*. Environmental Entomology. 2014;**43**(1):242-251. DOI: 10.1603/en13184

[40] Riddick EW, Barbosa P. Cry3A-intoxicated *Leptinotarsa* decemlineata (say) are palatable prey for *Lebia* grandis Hentz. Journal of Entomological Science. 2000;**35**:342-346

[41] Garcia-Alonso M, Jacobs E, Raybould A, Nickson TE, Sowig P, Willekens H, et al. A tiered system for assessing the risk of genetically modified plants to non-target organisms. Environmental Biosafety Research. 2006;**5**:57-65

[42] Duan JJ, Teixeira D, Huesing JE, Jiang CJ. Assessing the risk to non-target organisms from Bt corn resistant to corn rootworms (*Coleoptera*: *Chrysomelidae*): Tier-I testing with *Orius insidiosus* (*Heteroptera*: *Anthocoridae*). Environmental Entomology. 2008;**37**:838-844

[43] Naranjo SE. Long-term assessment of the effects of transgenic Bt cotton on the abundance of nontarget arthropod natural enemies. Environmental Entomology. 2005;**34**:1193-1210

[44] Cui JJ, Xia JY. Effects of transgenic Bt-cotton on development and reproduction of cotton bollworm. Acta Agriculturae Universitatis Henanensis. 1999;**33**:20-24

[45] Lu J, Liu YB, Shelton AM. Laboratory evaluations of a wild crucifer *Barbarea vulgaris* as a

management tool for diamondback moth. Bulletin of Entomological Research. 2004;**94**:509-516

[46] Baur ME, Boethel DJ. Effect of Bt-cotton expressing Cry1A(c) on the survival and fecundity of two hymenopteran parasitoids (*Braconidae*, *Encyrtidae*) in the laboratory. Biological Control. 2003;**26**:325-332

[47] Schuler TH, Potting RPJ, Denholm I, Poppy GM. Parasitoid behaviour and Bt plants. Nature. 1999;**400**:825-826

[48] Blumberg D, Navon A, Keren S, Goldenberg S, Ferkovich SM. Interactions among *Helicoverpa armigera* (*Lepidoptera*: *Noctuidae*), its larval endoparasitoid *Micropilis croceipes* (*Hymenoptera*: *Braconidae*), and *Bacillus thurigiensis*. Journal of Economic Entomology. 1997;**90**:1181-1186

[49] Schuler TH, Denholm I, Clark SJ, Stewart CN, Poppy GM. Effects of Bt plants on the development and survival of the parasitoid Cotesia plutellae (*Hymenoptera*: *Braconidae*) in susceptible and Bt-resistant larvae of the diamondback moth, *Plutella xylostella* (*Lepidoptera*: *Plutellidae*). Journal of Insect Physiology. 2004;**50**:435-443

[50] Schnepf E et al. Bacillus thuringiensis and its pesticidal crystal proteins. Microbiology and Molecular Biology Reviews. 1998;**62**:775-806

[51] Ayra-Pardo C, Raymond B, Gulzar A, Rodríguez-Cabrera L, Morán Bertot I, Crickmore N, et al. Novel genetic factors involved in resistance to *Bacillus thuringiensis* in *Plutella xylostella*. Insect Molecular Biology. 2015;**24**:589-600

[52] Tabashnik BE, Cushing NL, Finson N, Johnson MW. Field development of resistance to *Bacillus thuringiensis* in diamondback moth (*Lepidoptera*: *Plutellidae*). Journal of Economic Entomology. 1990;**83**:1671-1676

[53] Jiang T, Wu S, Yang T, Zhu C, Gao C. Monitoring field populations of *Plutella xylostella* (*Lepidoptera*: *Plutellidae*) for resistance to eight insecticides in China. Florida Entomologist. 2015;**98**:65-73

[54] Li J, Wu J, Yu Z, Akhurst R. Resistance of *Plutella xylostella* (L.) to *Bacillus thuringiensis*. Journal of Huazhong Agricultural University. 1997;**17**:214-217

[55] Wang C, Wu S, Yang Y, Wu Y. Field-evolved resistance to Bt delta endotoxins and Bt formulation in {*Plutella xylostella*} from the southeastern coast region of China. Acta Entomologica Sinica. 2005;**49**:70-73

[56] Xia Y, Lu Y, Shen J, Gao X, Qiu H, Li J. Resistance monitoring for eight insecticides in *Plutella xylostella* in Central China. Crop Protection. 2014;**63**:131-137

[57] Obrist LB, Dutton A, Albajes R, Bigler F. Exposure of arthropod predators to Cry1Ab toxin in Bt maize fields. Ecological Entomology. 2006;**31**:1-12

[58] Burd AD, Bradley JR, Van Duyn JW Jr, Gould F. Resistance of bollworm, *Helicoverpa zea*, to CryIA(c) toxin. In: Dugger CP, Richter DA, editors. Proceedings, 2000 Beltwide Cotton Conferences; 4-8 January 2000; San Antonio, TX: National Cotton Council of America, Memphis, TN; 2000. pp. 923-926

[59] Morin S, Henderson S, Fabrick JA, Carrière Y, Dennehy TJ, et al. DNA-based detection of Bt resistance alleles in pink bollworm. Insect Biochemistry and Molecular Biology. 2003;**34**:1225-1233

[60] Wan P, Wu K, Huang M, Wu J. Seasonal pattern of infestation by pink bollworm *Pectinophora gossypiella* (Saunders) in field plots of Bt transgenic cotton in the Yangtze River

Valley of China. Crop Protection. 2004;**23**:463-467

[61] Tabashnik BE, Liu YB, Unnithan DC, Carrière Y, Dennehy TJ, et al. Shared genetic basis of resistance to Bt toxin Cry1Ac in independent strains of pink bollworm. Journal of Economic Entomology. 2004;**97**:721-726

[62] Fabrick J, Oppert C, Lorenzen MD, Morris K, Oppert B, et al. A novel *Tenebrio molitor* cadherin is a functional receptor for *Bacillus thuringiensis* Cry3Aa toxin. The Journal of Biological Chemistry. 2009;**284**:18401-18410

[63] Shelton AM, Wyman JA, Cushing NL, Apfelbeck K, Dennehy TJ, Mahr SER, et al. Insecticide resistance of diamondback moth (*Lepidoptera*: *Plutellidae*) in North America. Journal of Economic Entomology. 1993;**86**:11-19

[64] Feng J, Pandey RB, Berry RJ, Farmer BL, Naik RR, Heinz H. Adsorption mechanism of single amino acid and surfactant molecules to Au 111 surfaces in aqueous solution: Design rules for metal-binding molecules. Soft Matter. 2011;**7**(5):2113-2120

[65] Shelton AM, Nault BA. Dead-end trap cropping: A technique to improve management of the diamondback moth, *Plutella xylostella* (*Lepidoptera*: *Plutellidae*). Crop Protection. 2004;**23**:497-503

[66] Satpathy S, Shivalingaswamy T, Kumar A, Rai A, Rai M. Potentiality of Chinese cabbage (*Brassica rapa* subsp. Pekinensis) as a trap crop for diamondback moth (*Plutella xylostella*) management in cabbage. Indian Journal of Agricultural Sciences. 2010;**80**:238-241

[67] Badenes-Perez FR, Shelton AM, Nault BA. Evaluating trap crops for diamondback moth, *Plutella xylostella* (*Lepidoptera*: *Plutellidae*). Journal of Economic Entomology. 2004;**97**:1365-1372

[68] Furlong MJ, Wright DJ, Dosdall LM. Diamondback moth ecology and management: Problems, progress, and prospects. Annual Review of Entomology. 2013;**58**:517-541

[69] Ramachandran S, Buntin GD, All JN, Tabashnik BE, Raymer PL, Adang MJ, et al. Survival, development, and oviposition of resistant diamondback moth (*Lepidoptera*: *Plutellidae*) on transgenic canola producing a *Bacillus thuringiensis* toxin. Journal of Economic Entomology. 1998;**91**:1239-1244

[70] Kamble A, Koopmann B, von Tiedemann A. Induced resistance to *Verticillium longisporum* in *Brassica napus* by β-aminobutyric acid. Plant Pathology. 2012;**62**(3):552-561

[71] Prasad YG, Prabhakar M, Sreedevi G, Ramachandra Rao G, Venkateswarlu B. Effect of temperature on development, survival and reproduction of the mealybug, *Phenacoccus solenopsis* Tinsley (Hemiptera: Pseudococcidae) on cotton. Crop Protection. 2012;**39**:81-88

[72] Kato M, Itioka T, Sakai S, Momose K, Yamane S, Hamid AA, et al. Various population fluctuation patterns of light-attracted beetles in a tropical lowland dipterocarp forest in Sarawak. Population Ecology. 2000;**42**:97-104

[73] Raimondo S, Strazanac JS, Butler L. Comparison of sampling techniques used in studying *Lepidoptera* population dynamics. Environmental Entomology. 2004;**33**:418-425

[74] Saito T, Rushtapakornchai W, Vattanatangum A, Sinchaisri N. Yellow trap experiments. In: Report on Insect Toxicological Studies on Resistance to Insecticides and Integrated Control of Diamondback Moth; Bangkok, Thailand: Department of Agriculture; 1988. pp. 76-81

[75] Butenandt VA, Beckmann R, Stamm D, Hecker E. Uber den sexual lockstoff

des seidenspinners Bombyx mori . Reindarstellung und knostitution. Z. Naturforsch. 1959;**14**:283-284

[76] Baker TC. Use of pheromones in IPM. In: Radcliffe T, Hutchinson B, editors. Integrated Pest Management. Cambridge: Cambridge University Press; 2008. pp. 273-285

[77] Sharov AA, Leonard D, Liebhold AM, Clemens NS. Evaluation of preventive treatments in low-density gypsy moth populations using pheromone traps. Journal of Economic Entomology. 2002;**95**:1205-1215

[78] Shahid AA, Bano S, Khalid S, Samiullah TR, Bajwa KS, et al. Biosafety assessment of transgenic Bt cotton on model animals. Advancements in Life Sciences. 2016;**3**(3):97-108

Chapter 3

Abiotic Stress Tolerance in Cotton

Aamir Hassan, Muhammad Ijaz, Abdul Sattar, Ahmad Sher, Sami-Ullah, Iqra Rasheed, Muhammad Zain Saleem and Ijaz Hussain

Abstract

Cotton (*Gossypium hirsutum* L.) is a vital fiber crop that is being cultivated under diverse climatic conditions across the globe. The demand for cotton and its by-products is increasing day by day due to more consumption of this fiber in the textile industry and the utilization of cotton seed as a source of edible oil. However, the average seed cotton yield in the world is below that of the potential yield of cultivars. The factors responsible for low yield includes shortage of approved seed, pest and disease attack, weed infestation, unwise use of nutrients, and the incidence of abiotic stresses (including drought, heat, and salinity). Among these, the abiotic stresses are a single major factor, which is responsible for reducing the yield now and will affect the productivity of cotton in future. In this scenario, it is necessary to adopt ways to improve the tolerance of cotton against abiotic stresses. The strategies for improving tolerance against abiotic stresses may include the wise use of macro- and micronutrients, the use of osmoprotectants, the use of arbuscular mycorrhizal fungi, and the plant-growth promoting rhizobacteria.

Keywords: nutrient management, PGPRs, osmolytes, plant hormones, fiber

1. Introduction

Abiotic stresses are major limiting factors that affect the growth, yield, and development of cotton. It is a fiber crop. It is cultivated in many countries across the globe. Medicinal products, home stuff, and cloth products are being processed from cotton crop. The raw material for the textile industry and also human oil consumption requirement are fulfilled by this crop. Extreme temperature, salinity stress, and water depletion are the main abiotic stresses that are considered the primary factors, which limit the productivity of cotton. The worldwide reduction of cotton crop is 50% due to the abiotic stress [1].

For maximum yield of cotton crops, they require optimum growth conditions like other field crops. For example, a temperature of 27–32°C is preferred by cotton crop during the formation of boll. At ≥36°C [2], the major reduction in carbon fixation was found in cotton crop, and for optimum photosynthesis, the optimum temperature is ~33°C. Poor yield and growth of the plant are caused by the major impact of salinity and alkalinity. The water stress in cotton is caused by salt acting as an osmoticum.

Specific ion toxicity is also a major cause of low yield in this crop. Inequity of nutrients is also a major cause. Plant metabolism is affected by impairing the photosynthetic process and membrane thermostability due to high temperature management. At the higher temperatures, the protein may be denatured and the activity of enzyme is more sensitive. Due to the effect of drought stress, the cell growth is influenced by the decrease in turgor pressure in cotton crop. The carbohydrate metabolism as well as photosynthesis is influenced by the drought stress directly or indirectly. Any change in carbon uptake also changes the process of photosynthesis resulting in the decrease of boll maintenance of the cotton plant and also the area of leaf that is a response to the stress due to drought [3]. The fiber quality, yield, and growth of this crop are affected by different abiotic stresses. In this chapter, we have discussed the impacts of different abiotic stresses on cotton performance and have enlisted possible improvement in its performance through application of plant hormones (auxin, cytokinins, abscisic acid, brassinosteroids, ethylene, and gibberellins) and plant nutrients (macronutrients and micronutrients).

2. Salinity stress

Throughout the biosphere, the salinity stress has been the most important restrictions for the productivity of agriculture [4]. The cultivated area affected by the stress of salinity all over the world is 20% [5]. Because of decline in water uptake by emerging seeds, plant roots, photosynthesis, respiration, and protein synthesis, germination is reduced due to impact of salt stress. It also affects productivity and growth of the cotton crop [6]. In mitochondria and chloroplast, the undue accumulation and generation of reactive oxygen species [ROS; like superoxide anion (O^{-2}), the hydroxyl radicals (OH), and hydrogen peroxide (H_2O_2)] are a result of the effect of soil salinity stress [7, 8]. Excessive salts in soil affect negatively the productivity and growth of cotton [9]; however, cotton is one of the most salt-tolerant crops. Plants own a number of antioxidant enzymes such ascorbate peroxidase (APX), glutathione reductase (GR), and superoxide dismutase (SOD) for fortification against the damaging effect of ROS (e.g., superoxide anion (O^{2-})) [10]. Under the unfavorable condition, the osmolytes metabolize the function in cotton to produce sugar alcohol [11]. Glycinebetaine and Proline serve as scavengers of ROS and also well-suited protectants, osmolytes for the macromolecules under the condition of salt stress. With specific references to stress and photosynthesis metabolism for controlling the survival and productivity, little information is present on biochemical and physiological features of cotton under the salt stress conditions. At the persistent salinity of 17 dS m^{-1}, the yield reduction is 50%, and when the salinity is at threshold level, which is 7.7 dS m^{-1}, a notable decline in seed cotton yield occurs. In conclusion, soil salinity negatively impacts the cotton growth and yield by affecting the plant physiological and biochemical traits.

3. Drought stress

For all the agricultural commodities, the availability of water is a determining factor for the yield and growth of the cotton plants under stress situations. Increasing human demand for water availability and demands for water for agriculture purpose in increasing and changing climate condition are the main factors restraining accessibility to water for agriculture. The shorter plants with small number of nodes resulted due to the drought stress in the cotton plants during the squaring period. With the help of drought stress treatments, there would be highest yields of the cotton plants. Except the full application of irrigation, the fiber quality parameters were significantly improved. The poor fiber quality, lowest fruit retention, and lowest yield production at the flowering stages are more sensitive to drought stress. There is poor fiber quality and yield losses at the squaring as a result of stress due to drought [12]. The severity and timing of the drought determine what will be the effect of water stress on yield. Photosynthesis process is slowed down because of a decrease in the number and size of cotton leaves. Krieg [13] showed that water stress reduced crop growth rate.

The variability in genotype responses to drought stress in cotton has been reported [14]. Compared to drought tolerance, many morpho-physiological characters have been recommended as significant selection criteria for cotton crops. The distance from the first main lateral root to transition zone is increased due to drought stress in cotton, and also the increase in taproot weight, seedling vigor, the amount of lateral roots, and also the development of root system is rapid [15]. The temperature of canopy, the discrimination of carbon isotope, leaf water content, conductance of stomata, and rate of photosynthesis also reduce the rate of transpiration due to the effect of drought [16]. Cotton crop has taproot system. In cotton seedlings, a number of lateral roots are produced, which depends on the xylem poles for water absorption [17]. The amount of vascular bundles increases due to the increase of branching intensities of lateral roots in the cotton crop [18]. In cotton, decreasing leaf transpiration by stomatal transpiration (TRst) and cuticular transpiration (TRcu) is the important physiological indicator of water stress [19].

Stomatal conductance controls stomatal transpiration (TRst) under water stress conditions. Leaf surface characters like morphological structure and the thickness of the wax layer affect the cuticular transpiration (TRcu) [20]. Lewitt showed that stomatal closing can avoid drought in plants. Stomatal closing and opening are regulated by the help of guard cells. Overproduction of reactive oxygen species (e.g., superoxide and peroxide) is followed by drought stress. Inhibition of photosynthesis and cellular damage are a result of this. This process is known as oxidative stress and is a major cause of plant damage due to stresses of environment [21] in many crops. According to McMichael et al. [17], in the present cotton cultivars, genetic variability is low for many drought-tolerant characters. So, under high rainfall and humid situations, much of the current cultivars are opted. Potential sources of traits associated with drought tolerance are considered as primitive race stocks of upland cotton [22].

4. Heat stress

By virtue of its geographical position, the cotton belt of Pakistan is present in the area of high level of temperature. In the Kharif season, the temperature approaches 50°C. The water stress and high temperature increase the impact that reduces the yield or quality of fiber, and fewer plants per unit area are a result of the heat stress with the other environmental stresses [23]. It is estimated that the harmful effect of heat stress causes the cotton crop to achieve only about 25% of yield potential [24]. The effect of these stresses is location-specific, exhibiting variation in frequency, intensity, and duration. The environmental stresses are site-specific, exhibiting frequency variation, light intensity, and duration of light. It is the practical approach to estimate the responses of heat responses by field evaluation of cotton under high temperatures with appropriate irrigations [10]. The ability to screen for heat tolerance might be affected by the timing of heat stress. It has been suggested that the identification of relative cell injury level from leaf disks at high temperature is the screening technique for heat tolerance in plants [25]. Plant development rate is much increased at high temperature, which reduces the life period besides other detrimental effects like denaturing of membranous structures [26]. Lint yields and quality are negatively correlated with the high temperature [27, 28]. The first and foremost requirement is to identify the suitable stock(s) to be used in breeding in any crop improvement program [29]. It was reported that in most dry land cotton production areas seedling heat tolerance is essential. Under heat conditions, emerging cotton seedlings poorly develop root system and show burning effects on the leaves; particularly, the younger leaves are adversely affected [11]. When plants grown in pots are exposed to high air temperatures, the shoots and the roots are challenged with hot condition, and it was observed that optimum temperature for leaf area development was 26°C for cotton [30].

5. Waterlogging stress

In areas with poor drainage or level, due to the excess of drainage and rainfall, the soil surface becomes saturated with water and this state of land is called waterlogging. Every year, the land area of the world experienced by waterlogging is about 10% [31]. The following are the two conditions: one is anoxic (oxygen absent: energy gain by fermentation is the only condition) and the other is hypoxic (low oxygen concentration: mitochondrial respiration is reduced and the process of fermentation takes place) because the microbial activity and plant activity use maximum amount of oxygen. During the conditions when the soil is waterlogged, the physicochemical properties such as the redox potential and pH are strongly changed due to the lack of oxygen concentration [32]. The effect of waterlogging on the salt-containing soil is more than 50% and these soils are mostly used for high-value crops such as cotton [33]. There are many drawbacks of the consequences of waterlogging for the plants of cotton, which may include terminated growth and the death of root apices, and also, increasing nutrient patterns may also be changed. For the growth of cotton, a waterlogged environment is lethal because it stops exchanging of gas and also results in energy problems [34]. Through the process of waterlogging, yield formation and the growth of cotton are strongly affected. But also these processes are complicated and remain unclear. It is reported that the adoption of cotton to the waterlogged stress is very poor [35]. But the cotton crop is that type of species that has indeterminate growth habit and has the large ability to compensate after the effect of abiotic stress.

6. Improving abiotic stress tolerance in cotton

6.1 Plant hormones

6.1.1 Auxins

For the development of the body and for the life cycle of plants, auxins are essential. These hormones play a critical role in the coordination of behavioral process and also in the growth of the plant. These hormones are present in all parts of plants. For the different process, the amount of these hormones is also different; for example, the most dominating and effective auxin is indole acetic acid (IAA). For the growth of cotton plants in abiotic stress, the dynamic and environment-responsive pattern of this hormone distribution within the plants of cotton is a key factor for their growth. These are also very important for the development of plant organs such as leaves or flowers and for the environmental reaction under the abiotic stress. Through the plant body, the process of polar auxin transport is achieved by the complex and well-coordinated active movement of these hormones from cell to cell in the plant body. Indole-3-propionic acid, indole-3-butyric acid, phenylacetic acid, indole-3-acetic acid, and 4-chloroindole-3-acetic acid are the five naturally occurring auxins, which are endogenous in nature [36]. For the proper development of plant growth, these hormones are very essential and contribute to giving the shape to the organ. Plants would be merely successful heaps of similar cells without hormonal regulation and organization of auxin hormone. The development of primary growth poles and future buds are formed by the auxin application. The employment of auxin begins in the embryo of the plant, and for subsequent growth, the distribution of the hormone is directional under the abiotic stresses [37]. This hormone is very important for proper growth and development. Also, with the help of this hormone, fruit senescence is delayed. In cotton, auxin plays a small role in initiation of the flowering and for reproductive organ development. Under the abiotic stress condition, when there is low concentration of auxin hormone, the senescence of the flower is delayed. In cotton, the lower concentration of this hormone can inhibit the formation of ethylene and also higher concentration can disturb the synthesis of ethylene. In cotton plants under abiotic stress, the auxin hormone influences a different kind of process such as the developmental and physiological. Through the auxin hormone application under stress conditions, rapid alteration in the roots of cotton occurs. Under abiotic stresses, in the cotton plants, various signaling auxin components appear that mediate diverse physiological and developmental processes. The target of various auxin-signaling components might be the strategy of potential to enhance the tolerance in cotton plants under abiotic stresses.

6.1.2 Cytokinins

Cytokinins are naturally occurring type of plant hormones. Under the drought condition, with the help of that hormone, the production of cotton is increased under stress. This increases the cell division and growth. The growth of the plant's main stem and branches is motivated in cotton by these hormones. For the growth and yield of cotton, there are many commercially produced hormones available, which are applied under the stress condition. In the area where there is absence of water or no irrigation, through the application of these hormones, the growth is also improved under stress conditions. Half of the production of cotton from Asia is in arid high water-shortage areas. The 60–65% of the acreage in the area is dry and depends on the rainfall for the moisture of the soil in short growing season. There is more difficulty for the cotton plants to absorb the soil water because the young cotton plant seedlings have small root systems under stress conditions. In the young plant, the defense for the water is promoted by that hormone. Also for the absorption of the soil moisture, it helps to promote the plants to build a strong and deep root system. To prevent the loss of water under stress conditions, it stimulates the growth of protective wax on the surface of the plants. Under water-stressed conditions, it has been reported that the application of cytokinins increases the yield by 5–10%. The cytokinins can be applied in the early season when conducting normal weed management practices, and no extra work is involved for the grower. It should be applied at a relatively low concentration to cotton seeds or to cotton plants at an early stage of development. The developmental and various physiological processes in the cotton plants are done by cytokinins. The division of the cell in plants also increases under the abiotic stress [38].

Cytokinins have a vital function in seed and root development. This hormone also retarded fiber elongation at elevated concentration in ovule culture. Cotton fiber and seed yield were improved by slightly raising the level of endogenous cytokinins. This also decreases the expression of cytokinin dehydrogenase [39]. Plant hormones play a significant role during interaction with physiological and developmental 'switches' involved in fiber growth. Cytokines also help in cell elongation by loosening the cell wall and supplying structural materials under stress conditions. During this process, secondary cell wall deposition and increased cellulose formation are key roles of that hormone. The opposed effect of some hormones may act as a restraining factor for fiber cell development under the abiotic stress conditions. The exogenous application of plant growth regulators at a particular time may be helpful for the appropriate cell development. Little is known about how some of the cells are differentiated into lint (long fibers) and others into fuzz (short fibers) from the same ovule epidermis. Selective utilization of nutrients for elongation of long fibers is the main reason under the stress. When a number of cells differentiate into fiber, some substances from ovule epidermal cells are transferred into fuzz, which affects other cells to develop into full-length fibers, which is another important reason under the stress condition.

6.1.3 Abscisic acid

The role of the abscisic acid (ABA) in the fiber development is an inhibitor. The growth of the fiber is also decreased when using the ABA to unfertilized cultured ovules [40]. The inhibitory function of ABA is somewhat balanced in the presence of cytokinins, which inhibits fiber development in the absence of ABA. At the time of boll formation, the concentration of ABA is low and also decreases during the next 2 days [41].

It was found that the ABA level was higher in mature cotton fruits as compared to young healthy fruits [42]. It was concluded that the internal ABA level exhibited a reverse correlation with the rate of fiber elongation. Among the different cotton cultivars, it is shown that high internal ABA contents result in shorter fiber and the reverse relationship exists between ABA contents and fiber length. Dasani and Thaker [43] tested the fiber of different cultivars of cotton under stress condition. The function of the ABA is revealed in both in vitro and in vivo situations for the improvement of fiber. The inhibitory effect of ABA on fiber length was reduced due to the addition of growth promoters like naphthaleneacetic acid (NAA) and gibberellic acid (GA) along with ABA. From the results of in vivo and in vitro experiments, it can be concluded that ABA may be playing an inhibitory role in fiber elongation and is a positive indicator of the onset of cell wall thickening.

6.1.4 Brassinosteroids

Brassinosteroids are naturally occurring hormones with steroid chemistry and are found throughout the kingdom Plantae. They elicit growth stimulation at nanomolar concentrations. Brassinosteroids enhance cell elongation and affect cytoskeleton and cell wall structure.

It is stated that adding a minute concentration of brassinosteroid (brassinolide (BL)) to cultured cotton ovules increased cotton fiber elongation, while the use of brassinazole 2001 (BRZ) and also the inhibitor of BR biosynthesis retarded fiber length and ovule size [44]. The application of BR biosynthesis inhibitor (brassinazole 2001) hindered fiber initiation probably due to alteration in the differentiation of ovule epidermal cells into fibers. The exogenous application of BL increases the formation of fiber, while the application of BRZ reverses the effect [45]. BR signal transduction plays a role in determining cotton fiber length. Transgenic plants with altered brassinosteroid insensitive 1 (BRI1) expression produce fibers similar in length to wild-type plants. The thicker secondary wall with fiber is produced by the plants that overexpress BRI1. These are the changes in fiber cell growth correlated with changing in expression of cellulose formation gene in fiber development.

6.1.5 Ethylene

Ethylene biosynthesis is the most important pathway that is upregulated during cotton fiber cell elongation in accordance with recent physiology and gene expression analysis [46] under optimal and suboptimal conditions. During the 10–15 DPA (days post anthesis), the involvement of 1-aminocyclopropane-1-carboxylic acid oxidase 1–3 (ACO1–3) was predicted very effective for fiber growth elongation under the abiotic stress condition. The exogenous application of the ethylene inhibitor, 2-aminoethoxyvinyl glycine (AVG), inhibits the growth of fiber, and ethylene increased fiber cell expansion under the stress condition [45]. According to the results, under the stress condition, this hormone has a significant role in supporting cotton fiber growth and elongation. Additionally, ethylene might enhance cell elongation by escalating the expression of tubulin, sucrose synthase, and expansion genes [46]. Detection of ethylene in fibers proved that it affects fiber elongation.

Ethylene biosynthesis genes (ACO1–3) are expressed at fiber elongation stage. According to that, it may interact with BR and ROS signaling pathway. Experiments on cultured ovules have shown that exogenous application of ethylene ameliorate the problem of fiber elongation caused due to BR biosynthesis inhibition. The exogenous application of both ethylene and BR on cultured ovules triggered the expression of genes for biosynthesis of other phytohormones. This cross-talk between hormones and genes may regulate fiber development in both negative and positive perspectives [47].

6.1.6 Gibberellins

The combination of auxin and gibberellins has been found to increase the fiber growth in in vitro cultured ovules [48]. Under abiotic stress, the application of auxin and gibberellins from exogenous source is vital for fiber growth in unfertilized ovules [49]. Studies on gene expression also explored the role of gibberellins and auxin in fiber growth. In DNA microarray, a cupin super family protein was found to be upregulated in 10 DPA ovules [50]. Because the plants have tissue sensitivity to improve the crop yield and quality, the transgenic approach has increased the manipulation of the hormones' concentration [51]. At a molecular level, to improve the fiber length and micronaire value, much effort has been made by scientists. Also the increased fiber for lint percentage and elongation was observed in cotton crop [52]. The targeted expression of an IAA biosynthetic gene under floral binding protein promoter (FBP7) was also shown in several studies and amplified the endogenous IAA levels at the fiber initiation stage under the abiotic stress [53]. The main aim of cotton-producing countries is to improve the yield of crop. By developing the seed that gives more yield of fiber under abiotic stress conditions, this aim of high yield can be fulfilled. The development of plant hormones plays an important role for the maximum growth and development of the crop [54]. The exogenous application of GA_3 not only promotes the fiber length but also enhances the thickness of cell wall significantly. During abiotic stress, long length cotton fibers with thicker cell wall and increased dry weight per unit cell length were obtained.

6.2 Plant nutrients

6.2.1 Macronutrients

6.2.1.1 Nitrogen

Nitrogen is a significant constituent of nucleic acids and amino acids and is required in high concentrations to plants. Maximum yields are not obtained from optimum nitrogen supply in the absence of adequate water, and optimum water supply will also not give maximum yield in the absence of adequate nitrogen supply [55]. Cotton that grows in different moisture stress levels in sandy soil shows similar special interactive effects of nitrogen supply and drought stress. Nitrogen shows genetic variation, selection, and breeding of lineages that are more effective in their N uptake. It is the more efficient strategy in arid land than in temperate zone [56]. When salinity is not severe, the addition of nitrogen enhances the growth and yield of crops [57].

Nitrogen also plays a key role in the synthesis of chlorophyll and proteins as well as in cell division. But cotton production can also be improved by foliar application in salinity stress [58]. Root development, germination, senescence, respiration, cell death, disease resistance, and hormone responses in crops are also influenced by nitrogen application. During abiotic stress in cotton, nitrogen plays an important role to activate the antioxidant defense in cotton [59]. Therefore, when the supply of nitrogen is adequate, root restriction increases the root activity. It also increases the availability of photoassimilates to above-ground plant parts. Hence, with the application of nitrogen to cotton, shoot growth and the ratio of shoot and root are enhanced.

6.2.1.2 Phosphorus

Phosphorus (P) is an essential component of nucleic acids, phosphor-lipids, and adenosine triphosphate. It also plays an important role in the storage, energy transfer, and also transport of carbohydrate. The pH is high and soils are calcareous in arid areas. Under the drought stress condition, phosphorous application can improve the growth of cotton crop [13]. The foliar application of urea and diammonium phosphate is the main source of phosphorous for the improvement of growth and development of cotton crop [60–62]. Improvement of fiber in cotton crop under the stress conditions can be obtained by the foliar spray of phosphorous at the boll formation stage [63]. In addition, boll weight and seed cotton yield are increased under stress [64].

Phosphorous is constituent of cell nuclei, and it is essential for cell division and development of meristematic tissues [65]. Phosphorous also influences the formation of nucleic acid, protein, and lipids as well as photosynthesis. In biotic stress conditions, the application of phosphorous improves the quality parameters of cotton. Cotton shows positive and economical response to phosphorous application [66]. Hence, plant height, shoots, and roots in cotton plants in abiotic stress conditions are enhanced by the application of phosphorous.

Phosphorous is efficiently applied to soil by fertigation as compared to broadcast application. However, in abiotic stress conditions, cotton yield can be improved with adequate amount of phosphorous fertilizer application at appropriate time. The reduced canopy is the result of the unbalanced nutrients in soil from the improper input of nutrients. Therefore, under abiotic stress conditions, photosynthesis rate and the yield of the cotton are reduced [67].

In abiotic stress conditions, the rate of leaf expansion and photosynthesis per unit leaf area of cotton crop are reduced due to phosphorous deficiency [68]. Crop growth, nitrogen and potassium uptake, total chlorophyll content, and dry matter yield of cotton plant are significantly enhanced by phosphorous [69]. The application of phosphorous leads to increased phosphorous uptake and content in leaf, stem, and reproductive parts such as seeds [70]. Phosphorous has a stimulating effect on number of flower buds and bolls per plant as well as is essential for cell division. Plant height, number of sympodial branches, seed index, boll weight, and seed cotton yield vary in all cotton cultivars due to genotypic variation [71, 72].

Cotton is facing decline in yield and quality because of abiotic stresses. Several genes for genetic engineering have been made from the cloning technology such as those related to fiber development (cytokinin dehydrogenase), disease resistance

(PR-3 and PR-10), and stress responses (GbRLI)[3]. These genes play an important role in successfully generating transgenic cotton lines with greater abiotic stress tolerance [73].

6.2.1.3 Calcium

Calcium plays a vital role in maintaining the many physiological processes that impact both the growth of cotton plants and also the responses to environmental stress. All the biotic and abiotic stresses and damages are repaired and act as defense for the cotton plants by the processes of translocation and respiratory metabolism. Concentration of water and the movement of the solutes influence these processes. These processes are also influenced by the Ca^{2+} on the structure of membrane and on the function of stomata. The uptake of calcium is minimized under stress conditions as compared to other elements. Hence, the accumulation of calcium is decreased to small extent as compared to phosphorous and potassium and this accumulation was in the range of 40, 71, and 91% for phosphorous, potassium, and calcium, respectively, in dry conditions in the mature cotton crops. The direct application of calcium is an efficient method for increasing the fiber yield of cotton. The incidence of fungal pathogens is reduced leading to increase in yield, and several physiological disorders are minimized by the application of calcium salt.

6.2.1.4 Potassium

The optimal supply and the good source of potassium (K) are very critical for increasing the growth and yield of the cotton crop. With the help of stomatal cell, the turgor pressure and osmotic pressure are increased with the help of K under the drought stress condition [74]. Soil salinity problem widely affects all the agronomic and physiological parameters of the cotton crop. These effects were lowered by the optimal application of potassium fertilizers [75]. Potassium increases the uptake of other essential nutrients, so the productivity of cotton is badly affected through the low application of potassium [76]. With no application of potassium, the cotton yield and also yield-contributing factors and fiber quality will reduce [77]. It was suggested in a study that under drought stress, the application of potassium influences the physiological functions of cotton [78]. The two cultivars of cotton were planted in drought stress and well-watered conditions with three potassium rates (0, 150, and 300 K_2O kg/ha) and these plants were showing higher leaf water potential, stomatal conductance, photosynthesis rate, and the maximum and actual quantum yield of PSII. With the application of potassium, the cotton plants were showing lower lipid peroxidation, higher antioxidant enzyme activity, as well as increased proline accumulation as compared to nonapplication of potassium, and a significant relationship was observed between photosynthetic recovery and potassium application.

Maintaining surplus water pressure within the boll also decreases the incidence of disease and improves the water use efficiency and fiber quality with the application of potassium [79]. Potassium application in cotton is also believed to extend the absorption of nitrogen, which causes vigorous vegetative growth and seed cotton yield. Also, the use of potassium in cotton enhanced the metabolic activity and improved the staple length, tensile strength, and fiber length and decreased the amount of damaged fiber [14]. Several other studies have reported an improvement in yield of cotton seed and quality of fiber due to potassium input in cotton under optimal and suboptimal conditions [80–82]. Combined foliar application of magnesium in combination with potassium and nitrogen improved the seed cotton yield, fiber quality, leaf nitrogen, potassium and magnesium concentration, and water use efficiency of cotton. The improvement in fiber quality was also visible through improvement in fiber strength, staple length, and fiber uniformity index owing to combined foliar application of magnesium in combination with potassium and nitrogen in abiotic stress in cotton crops [83].

Potassium plays a role in maintaining nitrogen metabolism and osmotic adjustment to sustain growth in soil under drought conditions [78]. Cotton plants under drought stress with potassium application not only showed higher osmotic adjustment with accumulation of osmolytes as well as maintaining higher enzyme activity, soluble proteins, and chlorophyll content but also regulate the nitrogen metabolism as compared to the plants without K application [84].

6.2.1.5 Micronutrients

As the cropping intensity increases, magnesium (Mg) deficiency occurs more frequently. Deficiency symptoms of sulfur are associated with the decrease in atmospheric sulfur. The uptake of magnesium and sulfur nutrients is reduced in cotton crop under drought stress. It has severe consequences for S nutrition and crop production. The plants uptake micronutrients through the process of diffusion decline because there is low soil moisture [85]. Cotton crop needs smaller quantities of micronutrients. Therefore, the effect of drought stress on micronutrients (Mg and S) is not the same as for macronutrients (P and N). Due to drought stress, deficiency of boron occurs in cotton crop. Due to the accumulation of silicon under drought conditions, the growth of cotton is improved and silicon is accumulated due to the reduction in transpiration rate [86, 87]. The main factors of saline and sodic soil on which they depend for availability of micronutrients are solubility of the micronutrients, pH, and the nature of the binding sites on the organic- and inorganic-particle surfaces. Salinity stress also affects the concentration of micronutrients in cotton plants, and soil salinity levels are also influenced by the salinity stress [88]. Inorganic nutrients play a significant role in determining plants' resistance to drought or salinity. Hence, both growth and development of cotton plants are similarly influenced by drought and salinity.

7. Use of osmoprotectants

The accumulation of organic osmolytes has been reported in many plants under abiotic stresses. These include polyhydroxylic compounds and zwitterionic alkyl amines. The accumulation of osmolytes is widely discussed nowadays especially in cotton crops [89, 90].

Osmotically active solute is completed by the entry of water into the cell. This water provides sufficient concentrations for turgor pressure, which is necessary for the expansion of cells.

Cotton plants remain fit under stressful environmental conditions due to osmotic adjustment [91]. Therefore, high concentrations of several but not all compatible solutes protect the crop from oxidative damage. Their damage is reduced by scavenging free radicals in addition to their rules in preservation of osmotic equilibrium without disturbing macromolecule solvent relations.

The resistance against the oxidative stress of cotton has recently increased with the action of chloroplast accumulation of mannitol as well as consistent with high diffusion rate limited reactivity of hydroxyl radicals toward the most metabolic intermediates [92]. A significant role is played by the compatible solutes in terminating free radical chain reaction. The stress tolerance appears due to the critical element glycinebetaine in the cotton plants.

The growth of cotton plants is strongly influenced by the drought and saline environments with the osmoprotectants. Osmoprotectants are enormously proficient compatible solutes. The accumulation of glycinebetaine is induced and improves the tolerance to abiotic stress conditions [93]. The treatment of cotton seeds with the external application of glycinebetaine at increased the cotton seed yield by 18 and 22%, respectively. The growth and survival of extensive varieties of plants such as cotton crops are improved by the exogenous application of glycinebetaine.

Author details

Aamir Hassan, Muhammad Ijaz*, Abdul Sattar, Ahmad Sher, Sami-Ullah,
Iqra Rasheed, Muhammad Zain Saleem and Ijaz Hussain
Bahauddin Zakariya University, Layyah, Pakistan

*Address all correspondence to: muhammad.ijaz@bzu.edu.pk

References

[1] Bita C, Gerats T. Plant tolerance to high temperature in a changing environment: scientific fundamentals and production of heat stress-tolerant crops. Frontiers in Plant Science. 31 Jul 2013;**4**:273

[2] Bibi AC, Oosterhuis DM, Gonias ED. Exogenous application of putrescine ameliorates the effect of high temperature in *Gossypium hirsutum* L. flowers and fruit development. Journal of Agronomy and Crop Science. Jun 2010;**196**(3):205-211

[3] Leakey AD, Ainsworth EA, Bernacchi CJ, Rogers A, Long SP, Ort DR. Elevated CO_2 effects on plant carbon, nitrogen, and water relations: Six important lessons from FACE. Journal of Experimental Botany. 28 Apr 2009;**60**(10):2859-2876

[4] McGrath JM, Lobell DB. Reduction of transpiration and altered nutrient allocation contribute to nutrient decline of crops grown in elevated CO_2 concentrations. Plant, Cell & Environment. Mar 2013;**36**(3):697-705

[5] Liu J, Zhu JK. Proline accumulation and salt-stress-induced gene expression in a salt-hypersensitive mutant of Arabidopsis. Plant Physiology. 1 Jun 1997;**114**(2):591-596

[6] Wang SL, Heisey PW, Huffman WE, Fuglie KO. Public R&D, private R&D, and US agricultural productivity growth: Dynamic and long-run relationships. American Journal of Agricultural Economics. 14 Jun 2013;**95**(5):1287-1293

[7] Miller G, Shulaev V, Mittler R. Reactive oxygen signaling and abiotic stress. Physiologia Plantarum. Jul 2008;**133**(3):481-489

[8] Masood A, Shah NA, Zeeshan M, Abraham G. Differential response of antioxidant enzymes to salinity stress in two varieties of Azolla (*Azolla pinnata* and *Azolla filiculoides*). Environmental and Experimental Botany. 1 Dec 2006;**58**(1-3):216-222

[9] Qadir M, Shams M. Some agronomic and physiological aspects of salt tolerance in cotton (*Gossypium hirsutum* L.). Journal of Agronomy and Crop Science. Oct 1997;**179**(2):101-106

[10] Asada M, Kanaya T, Nakatsuji M, Kamaguchi R, Iwamoto Y. Method for Storage of Seeds. United States Patent Application US 13/378,792. Morishita Jintan Co Ltd, Niigata University, assignee; 31 May 2012

[11] Ashraf MF, Foolad M. Roles of glycine betaine and proline in improving plant abiotic stress resistance. Environmental and Experimental Botany. 1 Mar 2007;**59**(2):206-216

[12] Lv F, Liu J, Ma Y, Chen J, Wang Y, Chen B, et al. Effect of shading on cotton yield and quality on different fruiting branches. Crop Science. 2013;**53**(6):2670-2678

[13] Ackerson RC. Osmoregulation in cotton in response to water stress: III. Effects of phosphorus fertility. Plant Physiology. 1985;**77**(2):309-312

[14] Ali H, Afzal MN, Muhammad D. Effect of sowing dates and plant spacing on growth and dry matter partitioning in cotton (*Gossypium hirsutum* L.). Pakistan Journal of Botany. 2009;**41**(5):2145-2155

[15] Cook D, Herbert A, Akin DS, Reed J. Biology, crop injury, and management of thrips (Thysanoptera: Thripidae) infesting cotton seedlings in the United States. Journal of Integrated Pest Management. 1 Oct 2011;**2**(2):B1-B9

[16] Nepomuceno AL, Stewart JM, Oosterhuis D, Turley R, Neumaier M,

Farias JR. Isolation of a cotton NADP (H) oxidase homologue induced by drought stress. Pesquisa Agropecuária Brasileira. Jul 2000;**35**(7):1407-1416

[17] McMichael BL, Quisenberry JE, Upchruch DR. Lateral root development in exotic cottons. Environmental and Experimental Botany. 1 Oct 1987;**27**(4):499-502

[18] Longenberger PS, Smith CW, Duke SE, McMichael BL. Evaluation of chlorophyll fluorescence as a tool for the identification of drought tolerance in upland cotton. Euphytica. 1 Mar 2009;**166**(1):25

[19] Eickmeier WG, Casper C, Osmond CB. Chlorophyll fluorescence in the resurrection plant *Selaginella lepidophylla* (Hook. & Grev.) Spring during high-light and desiccation stress, and evidence for zeaxanthin-associated photoprotection. Planta. 1 Jan 1993;**189**(1):30-38

[20] Richards RA, Rebetzke GJ, Condon AG, Van Herwaarden AF. Breeding opportunities for increasing the efficiency of water use and crop yield in temperate cereals. Crop Science. 1 Jan 2002;**42**(1):111-121

[21] Sunkar R, Bartels D, Kirch HH. Overexpression of a stress-inducible aldehyde dehydrogenase gene from *Arabidopsis thaliana* in transgenic plants improves stress tolerance. The Plant Journal. Aug 2003;**35**(4):452-464

[22] Basal H, Smith CW, Thaxton PS, Hemphill JK. Seedling drought tolerance in upland cotton. Crop Science. 1 Mar 2005;**45**(2):766-771

[23] Rahman HU. Number and weight of cotton lint fibres: Variation due to high temperatures in the field. Australian Journal of Agricultural Research. 13 Jun 2006;**57**(5):583-590

[24] Burner DM, MacKown CT. Nitrogen effects on herbage nitrogen use and nutritive value in a meadow and loblolly pine alley. Crop Science. 1 May 2006;**46**(3):1149-1155

[25] Toews MD, Tubbs RS, Wann DQ, Sullivan D. Thrips (Thysanoptera: Thripidae) mitigation in seedling cotton using strip tillage and winter cover crops. Pest Management Science. Oct 2010;**66**(10):1089-1095

[26] Boex-Fontvieille E, Davanture M, Jossier M, Zivy M, Hodges M, Tcherkez G. Photosynthetic activity influences cellulose biosynthesis and phosphorylation of proteins involved therein in Arabidopsis leaves. Journal of Experimental Botany. 19 Jul 2014;**65**(17):4997-5010

[27] Reddy KR, Hodges HF, McKinion JM. A comparison of scenarios for the effect of global climate change on cotton growth and yield. Functional Plant Biology. 1997;**24**(6):707-713

[28] Singh SK, Badgujar G, Reddy VR, Fleisher DH, Bunce JA. Carbon dioxide diffusion across stomata and mesophyll and photo-biochemical processes as affected by growth CO_2 and phosphorus nutrition in cotton. Journal of Plant Physiology. 15 Jun 2013;**170**(9):801-813

[29] Burke JJ. Evaluation of source leaf responses to water-deficit stresses in cotton using a novel stress bioassay. Plant Physiology. 1 Jan 2007;**143**(1):108-121

[30] Volkov VA, Bulushev BV, Ageev AA. Determination of the capillary size and contact angle of fibers from the kinetics of liquid rise along the vertical samples of fabrics and nonwoven materials. Colloid Journal. 1 Jul 2003;**65**(4):523-525

[31] Volkov SV, Grinchenko OS, Sviridova TV. The effects of weather and climate changes on the timing of autumn migration of the common crane (*Grus grus*) in the north of

Moscow region. Biology Bulletin. 1 Dec 2016;**43**(9):1203-1211

[32] Pezeshki SR, DeLaune RD, Patrick WH Jr. Flooding and saltwater intrusion: Potential effects on survival and productivity of wetland forests along the US Gulf Coast. Forest Ecology and Management. 1 Jun 1990;**33**:287-301

[33] Jackson MB, Drew MC, Giffard SC. Effects of applying ethylene to the root system of *Zea mays* on growth and nutrient concentration in relation to flooding tolerance. Physiologia Plantarum. May 1981;**52**(1):23-28

[34] Voesenek LA, Bailey-Serres J. Flooding tolerance: O_2 sensing and survival strategies. Current Opinion in Plant Biology. 1 Oct 2013;**16**(5):647-653

[35] Brodrick R, Bange MP, Milroy SP, Hammer GL. Physiological determinants of high yielding ultra-narrow row cotton: Biomass accumulation and partitioning. Field Crops Research. 12 Aug 2012;**134**:122-129

[36] Alvarez S, Marsh EL, Schroeder SG, Schachtman DP. Metabolomic and proteomic changes in the xylem sap of maize under drought. Plant, Cell & Environment. Mar 2008;**31**(3):325-340

[37] Schachtman DP, Goodger JQ. Chemical root to shoot signaling under drought. Trends in Plant Science. 1 Jun 2008 ;**13**(6):281-287

[38] Rupp HM, Frank M, Werner T, Strnad M, Schmülling T. Increased steady state mRNA levels of the STM and KNAT1 homeobox genes in cytokinin overproducing *Arabidopsis thaliana* indicate a role for cytokinins in the shoot apical meristem. The Plant Journal. Jun 1999;**18**(5):557-563

[39] Zhao J, Bai W, Zeng Q, Song S, Zhang M, Li X, et al. Moderately enhancing cytokinin level by down-regulation of GhCKX expression in cotton concurrently increases fiber and seed yield. Molecular Breeding. 2015;**35**(2):60

[40] Haigler CH, Zhang D, Wilkerson CG. Biotechnological improvement of cotton fibre maturity. Physiologia Plantarum. Jul 2005;**124**(3):285-294

[41] Haigler CH, Betancur L, Stiff MR, Tuttle JR. Cotton fiber: A powerful single-cell model for cell wall and cellulose research. Frontiers in Plant Science. 21 May 2012;**3**:104

[42] Gokani SJ, Thaker VS. Accumulation of abscisic acid in cotton fibre and seed of normal and abnormal bolls. The Journal of Agricultural Science. Dec 2001;**137**(4):445-451

[43] Dasani SH, Thaker VS. Role of abscisic acid in cotton fiber development. Russian Journal of Plant Physiology. 1 Jan 2006;**53**(1):62-67

[44] Sun Y, Veerabomma S, Abdel-Mageed HA, Fokar M, Asami T, Yoshida S, et al. Brassinosteroid regulates fiber development on cultured cotton ovules. Plant and Cell Physiology. 1 Aug 2005;**46**(8):1384-1391

[45] Shi YH, Zhu SW, Mao XZ, Feng JX, Qin YM, Zhang L, et al. Transcriptome profiling, molecular biological, and physiological studies reveal a major role for ethylene in cotton fiber cell elongation. The Plant Cell. 1 Mar 2006;**18**(3):651-664

[46] Shi R, Wang JP, Lin YC, Li Q, Sun YH, Chen H, et al. Tissue and cell-type co-expression networks of transcription factors and wood component genes in *Populus trichocarpa*. Planta. 1 May 2017;**245**(5):927-938

[47] Stiff MR, Haigler CH. Recent Advances in Cotton Fiber Development. Flowering and Fruiting in Cotton. Tennessee: The Cotton Foundation; 2012. pp. 163-192

[48] Seagull RW, Giavalis S, Seagull R, Giavalis S. Molecular biology and physiology pre-and post-anthesis application of exogenous hormones alters fiber production in *Gossypium hirsutum* L. cultivar maxxa GTO. Journal of Cotton Science. 2004;**8**:105-111

[49] Beasley CA, Ting IP. Effects of plant growth substances on in vitro fiber development from unfertilized cotton ovules. American Journal of Botany. Feb 1974;**61**(2):188-194

[50] Ji SJ, Lu YC, Feng JX, Wei G, Li J, Shi YH, et al. Isolation and analyses of genes preferentially expressed during early cotton fiber development by subtractive PCR and cDNA array. Nucleic Acids Research. 2003;**31**:2534-2543

[51] Ruan YL, Xu SM, White R, Furbank RT. Genotypic and developmental evidence for the role of plasmodesmatal regulation in cotton fiber elongation mediated by callose turnover. Cell Biology and Signal Transduction. 2004;**136**(4):4104-4113. DOI: 10.1104/pp.104.051540

[52] Xiao YH, Yan Q, Ding H, Luo M, Hou L, Zhang M, et al. Transcriptome and biochemical analyses revealed a detailed proanthocyanidin biosynthesis pathway in brown cotton fiber. PLoS One. 21 Jan 2014;**9**(1):e86344

[53] Yang Z, Zhang C, Yang X, Liu K, Wu Z, Zhang X, et al. PAG1, a cotton brassinosteroid catabolism gene, modulates fiber elongation. New Phytologist. 2014;**203**(2):437-448

[54] Luo M, Xiao Y, Li X, Lu X, Deng W, Li D, et al. GhDET2, a steroid 5α-reductase, plays an important role in cotton fiber cell initiation and elongation. The Plant Journal. Aug 2007;**51**(3):419-430

[55] Mengel R, Bacher M, Flores-de-Jacoby L. Interactions between stress, interleukin-1β, interleukin-6 and cortisol in periodontally diseased patients. Journal of Clinical Periodontology. Nov 2002;**29**(11):1012-1022

[56] Cairns JE, Crossa J, Zaidi PH, Grudloyma P, Sanchez C, Araus JL, et al. Identification of drought, heat, and combined drought and heat tolerant donors in maize. Crop Science. 2013;**53**(4):1335-1346

[57] Papadopoulos CE, Lazaridou A, Koutsoumba A, Kokkinos N, Christoforidis A, Nikolaou N. Optimization of cotton seed biodiesel quality (critical properties) through modification of its FAME composition by highly selective homogeneous hydrogenation. Bioresource Technology. 1 Mar 2010;**101**(6):1812-1819

[58] Chen W, Hou Z, Wu L, Liang Y, Wei C. Effects of salinity and nitrogen on cotton growth in arid environment. Plant and Soil. 1 Jan 2010;**326**(1-2):61-73

[59] Floryszak-Wieczorek J, Arasimowicz M, Milczarek G, Jelen H, Jackowiak H. Only an early nitric oxide burst and the following wave of secondary nitric oxide generation enhanced effective defence responses of pelargonium to a necrotrophic pathogen. New Phytologist. Sep 2007;**175**(4):718-730

[60] Ravindra S, Mohan YM, Reddy NN, Raju KM. Fabrication of antibacterial cotton fibres loaded with silver nanoparticles via "Green Approach". Colloids and Surfaces A: Physicochemical and Engineering Aspects. 5 Sep 2010;**367**(1-3):31-40

[61] Shahid MA, Pervez MA, Balal RM, Ahmad R, Ayyub CM, Abbas T, et al. Salt stress effects on some morphological and physiological characteristics of okra (*Abelmoschus esculentus* L.). Soil and Environment. 1 Jun 2011;**30**(1)

[62] Jabran K, Mahajan G, Sardana V, Chauhan BS. Allelopathy for weed control in agricultural systems. Crop Protection. 1 Jun 2015;**72**:57-65

[63] Singh R, Sharma RR, Tyagi SK. Pre-harvest foliar application of calcium and boron influences physiological disorders, fruit yield and quality of strawberry (Fragaria × ananassa Duch.). Scientia Horticulturae. 26 Mar 2007;**112**(2):215-220

[64] Rajakumar D, Gurumurthy S. Effect of plant density and nutrient spray on the yield attributes and yield of direct sown and polybag seedling planted hybrid cotton. Agricultural Science Digest. 2008;**28**(3):174-177

[65] Sawan ZM, Fahmy AH, Yousef SE. Effect of potassium, zinc and phosphorus on seed yield, seed viability and seedling vigor of cotton (*Gossypium barbadense* L.). Archives of Agronomy and Soil Science. 1 Feb 2011;**57**(1):75-90

[66] Gill SS, Tuteja N. Reactive oxygen species and antioxidant machinery in abiotic stress tolerance in crop plants. Plant Physiology and Biochemistry. 1 Dec 2010;**48**(12):909-930

[67] Jiang Y, Yang B, Harris NS, Deyholos MK. Comparative proteomic analysis of NaCl stress-responsive proteins in Arabidopsis roots. Journal of Experimental Botany. 1 Oct 2007;**58**(13):3591-3607

[68] Casadevall R, Rodriguez RE, Debernardi JM, Palatnik JF, Casati P. Repression of growth regulating factors by the microRNA396 inhibits cell proliferation by UV-B radiation in Arabidopsis leaves. The Plant Cell. 1 Sep 2013;**25**(9):3570-3583

[69] Sawan ZM, Mahmoud MH, El-Guibali AH. Influence of potassium fertilization and foliar application of zinc and phosphorus on growth, yield components, yield and fiber properties of Egyptian cotton (*Gossypium barbadense* L.). Journal of Plant Ecology. 1 Dec 2008;**1**(4):259-270

[70] Deshpande AN, Masram RS, Kamble BM. Effect of fertilizer levels on nutrient availability and yield of cotton on Vertisol at Rahuri, District Ahemadnagar, India. Journal of Applied and Natural Science. 1 Dec 2014;**6**(2):534-540

[71] Ghoneim AM, Gewaily EE, Osman MM. Effects of nitrogen levels on growth, yield and nitrogen use efficiency of some newly released Egyptian rice genotypes. Open Agriculture. 2018;**3**(1):310-318.s

[72] Baloch MJ, Khan NU, Rajput MA, Jatoi WA, Gul S, Rind IH, et al. Yield related morphological measures of short duration cotton genotypes. Journal of Animal and Plant Sciences. 1 Aug 2014;**24**(4):1198-1211

[73] Zimmermann MR, Mithöfer A, Will T, Felle HH, Furch AC. Herbivore-triggered electrophysiological reactions: Candidates for systemic signals in higher plants and the challenge of their identification. Plant Physiology. 2016;**170**(4):2407-2419

[74] Pervez H, Ashraf M, Makhdum MI. Effects of potassium rates and sources on fiber quality parameters in four cultivars of cotton grown in aridisols. Journal of Plant Nutrition. 2 Jan 2005;**27**(12):2235-2257

[75] Teotia P, Kumar V, Kumar M, Shrivastava N, Varma A. Rhizosphere microbes: Potassium solubilization and crop productivity–Present and future aspects. In: Potassium Solubilizing Microorganisms for Sustainable Agriculture. New Delhi: Springer; 2016. pp. 315-325

[76] Mullins GL, Burmester CH, Reeves DW. Cotton response to in-row subsoiling and potassium fertilizer

placement in Alabama. Soil and Tillage Research. 1 Jan 1997;**40**(3-4):145-154

[77] Bradow JM, Davidonis GH. Quantitation of fiber quality and the cotton production-processing interface: A physiologist's perspective. Journal of Cotton Science. May 2000;**4**(1):34-64

[78] Zahoor R, Zhao W, Abid M, Dong H, Zhou Z. Potassium application regulates nitrogen metabolism and osmotic adjustment in cotton (*Gossypium hirsutum* L.) functional leaf under drought stress. Journal of Plant Physiology. 2017;**215**:30-38

[79] Lin ZX, He D, Zhang XL, Nie Y, Guo X, Feng C, et al. Linkage map construction and mapping QTL for cotton fibre quality using SRAP, SSR and RAPD. Plant Breeding. Apr 2005;**124**(2):180-187

[80] Pervez H, Ashraf M, Makhdum MI. Influence of potassium nutrition on gas exchange characteristics and water relations in cotton (*Gossypium hirsutum* L.). Photosynthetica. 1 Jun 2004;**42**(2):251-255

[81] Pettigrew WT. Potassium influences on yield and quality production for maize, wheat, soybean and cotton. Physiologia Plantarum. Aug 2008;**133**(4):670-681

[82] Tsialtas IT, Shabala S, Baxevanos D, Matsi T. Effect of potassium fertilization on leaf physiology, fiber yield and quality in cotton (*Gossypium hirsutum* L.) under irrigated Mediterranean conditions. Field Crops Research. 1 Jul 2016;**193**:94-103

[83] Chen W, Feng C, Guo W, Shi D, Yang C. Comparative effects of osmotic-, salt- and alkali stress on growth, photosynthesis, and osmotic adjustment of cotton plants. Photosynthetica. 1 Sep 2011;**49**(3):417

[84] Wang H, Chen Y, Xu B, Hu W, Snider JL, Meng Y, et al. Long-term exposure to slightly elevated air temperature alleviates the negative impacts of short term waterlogging stress by altering nitrogen metabolism in cotton leaves. Plant Physiology and Biochemistry. 1 Feb 2018;**123**:242-251

[85] Irshad MU, Gill MA, Aziz TA, Ahmed I. Growth response of cotton cultivars to zinc deficiency stress in chelator-buffered nutrient solution. Pakistan Journal of Botany. 1 Jun 2004;**36**(2):373-380

[86] Läuchli A, Epstein E. Plant responses to saline and sodic conditions. Agricultural Salinity Assessment and Management. 1990;**71**:113-137

[87] Ma BL, Wu TY, Tremblay N, Deen W, McLaughlin NB, Morrison MJ, et al. On-farm assessment of the amount and timing of nitrogen fertilizer on ammonia volatilization. Agronomy Journal. 1 Jan 2010;**102**(1):134-144

[88] Oertli JJ. Nutrient management under water and salinity stress. In: Proceedings of the Symposium on Nutrient Management for Sustained Productivity. Ludhiana, India: Depart. Soils Punjab Agric. Unver.; 1991. pp. 138-165

[89] Bartels D, Phillips J. Drought stress tolerance. In: Genetic Modification of Plants. Berlin, Heidelberg: Springer; 2010. pp. 139-157

[90] Naranjo MA, Forment J, RoldÁN M, Serrano R, Vicente O. Overexpression of *Arabidopsis thaliana* LTL1, a salt-induced gene encoding a GDSL-motif lipase, increases salt tolerance in yeast and transgenic plants. Plant, Cell & Environment. Oct 2006;**29**(10):1890-1900

[91] Boyer JS, James RA, Munns R, Condon TA, Passioura JB. Osmotic adjustment leads to anomalously low estimates of relative water content in wheat and barley. Functional Plant Biology. 19 Dec 2008;**35**(11):1172-1182

[92] Lang J, Hu J, Ran W, Xu Y, Shen Q. Control of cotton Verticillium wilt and fungal diversity of rhizosphere soils by bio-organic fertilizer. Biology and Fertility of Soils. 1 Feb 2012;**48**(2):191-203

[93] Zhang K, Wang J, Lian L, Fan W, Guo N, Lv S. Increased chilling tolerance following transfer of a betA gene enhancing glycinebetaine synthesis in cotton (*Gossypium hirsutum* L.). Plant Molecular Biology Reporter. 2012;**30**(5):1158-1171

Chapter 4

Bioenergy Recovery from Cotton Stalk

Rafat Al Afif, Christoph Pfeifer and Tobias Pröll

Abstract

Cotton stalk (CS) plant residue left in the field following harvest must be buried or burned to prevent it from serving as an overwintering site for insects such as the pink bollworm (PBW). This pest incurs economic costs and detrimental environmental effects. However, CS contains lignin and carbohydrates, like cellulose and hemicelluloses, which can be converted into a variety of usable forms of energy. Thermochemical or biochemical processes are considered technologically advantageous solutions. This chapter reviews potential energy generation from cotton stalks through combustion, hydrothermal carbonization, pyrolysis, fermentation, and anaerobic digestion technologies, focusing on the most relevant technologies and on the properties of the different products. The chapter is concluded with some comments on the future potential of these processes.

Keywords: cotton stalk, thermochemical, biochemical, bioenergy

1. Introduction

Worldwide energy demand and greenhouse gas (GHG) emissions are predicted to increase by 70 and 60%, respectively, between 2011 and 2050 according to the International Energy Agency (IEA) [1]. An increase in GHG emissions is unequivocally the largest anthropogenic contributor to exacerbating climate change [2]. Currently, the majority of energy is derived from fossil fuels. As reported in 2017, it is estimated that if consumption of fossil fuels persists at 2016 levels, reserves of coal, gas, and oil will last only 153, 52.5, and 50.6 more years, respectively [3]. Therefore, other forms of energy such as biomass have significant potential to offset traditional energy sources [4]. Biomass, as a zero CO_2 emission fuel, can offer one solution in the reduction of CO_2 atmosphere content. In 2016 renewable energy accounted for 18.2% of the 576 exajoules (EJ) of total primary energy supply (TPES), of which 13% came from biomass [1, 5]. Biomass provided 46.4 EJ of TPES in 2016, and expert scientific analysis predicts that by 2050 the bioenergy share of TPES could reach 100–300 EJ per year (year^{-1}) with the highest theoretical share proposed at 500 EJ year^{-1} [5, 6]. Although renewable energy makes up only a small percentage of current TPES, it has the theoretical potential to provide all of the human energy requirements on earth [7]. By 2035, biofuels could realistically provide at least a quarter of the estimated world's TPES of 623 EJ. To increase the proportions of renewables in the TPES, innovative feedstocks or inputs are required [8]. A significant source of biomass for renewable energy is available globally in the form of agricultural waste. Agricultural wastes pose expensive and challenging

 IntechOpen

issues for crop producers. With exception to the fraction of residues tilled back into the soil to increase soil organic carbon (SOC) content and enhance other soil physical characteristics, many of these wastes have little to negative value, and knowledge of revenue streams are sparse [9, 10]. For example, cotton biomass waste is an abundant and available waste from agricultural production at a high estimate of roughly 50 million tons annually [11]. Similarly, other crops produce even more abundant waste, such as rice husks which sum up to 822 million tons of waste with no real end of use application [12].

It has been reported that cotton residues left during harvest are carriers of the pest; therefore, adequate disposal of these residues is necessary [13]. However, it is worth considering that one of the major complications of cotton production is the management of the pink bollworm (PBW) (*Pectinophora gossypiella*). It is considered one of the most detrimental cotton pests because of its hardiness to insecticides [14]. PBW's life cycle consists of four stages: egg, larva, pupa, and adult. During the first stage, females lay 200–500 tiny eggs in single or small groups of 5–10 each on cotton plants which hatch 3–4 days later. During the second and most destructive stage, the larvae bore into the bolls to grow before cotton boll blossoming occurs. Here the larvae feed on seeds for 12–15 days where they mature to 12 mm long as a fully developed larva. The most significant damage occurs to the seed and lint. Before pupation, the larva experiences diapause during the winter for 2–4 months in which they do not feed or move. They may be found in bolls, in stems, or in the soil in which they are safe in a silken cocoon until spring. During the pupation stage, spring conditions cause the larva to drop to the soil beneath the cotton plants where they pupate; the pupa is roughly 7 mm long and brown, and the pupal period is between 7 and 8 days. During the adult stage in spring, first-generation adults develop from the pupae and are gray brown small moth which mate and lay eggs. In the summer, larvae from the previous generation fall to the soil, pupate, and emerge as second-generation moths, completing the life cycle. The entire cycle from egg to egg takes roughly 32 days, and the PBW can persist, on average, for up to six summer cycles [15, 16]. This pest is distributed globally where cotton is grown and is considered the key cotton pest. Its main effect on cotton crops is preventing flowering buds to open, shedding of the fruit, seed loss, and damage to lint. Trials in the USA have shown that the potential loss of harvest without control was 61%, whereas losses of 9% were estimated when the pest was controlled through insecticide application. In 1998, the total US crop yield of cotton was reduced by 2.7%, while in Egypt it is estimated that the PBW causes losses of about 10–20% of cotton crop annually [13, 15]. In 2014, it was reported that the PBW had been eradicated from California, Arizona, New Mexico, and Texas in the USA as well as Chihuahua in Northern Mexico. The eradication is attributed to a combination of insecticides and genetic modification of the cotton crop as well as releases of sterile PBW throughout the region [17]. In countries without robust pest management strategies, the most common method of PBW prevention is through burning the residues in the field or by shredding and plowing the residues to a depth of 6 inches into the soil, the latter of which is time and energy intensive [11, 15]. In light of the global challenges associated with cotton agricultural residuals, a promising method of cotton waste disposal is through their utilization as an energy source.

Studies indicate that undebarked cotton stalks are unsuitable for the production of fine paper and dissolving pulps [18, 19]. Furthermore, cotton stalks and other agricultural residues are unsuitable for hardboard and particle board due to their high water absorption and thickness swelling (deteriorated dimensional stability) [20, 21].

In contrast, the usage of cotton waste as an energy feedstock has become a subject of numerous studies in recent years [22–24]. Researchers generally focused on the production of biogas, ethanol, and the production of fuel pellets or briquettes. Several studies on the subject of cotton waste pyrolysis indicate that pyrolysis of cotton stalks is deemed to have potential as one of the technological solutions for its management.

The purpose of the study is to review current bioenergy conversion technologies and to provide quantitative data and interpretation of the heating value, proximate and elemental analysis, and product yields specific to bioenergy recovery from CS. The hypothesis is that resulting data will be consistent with past research proving that CS residues have a high potential for use as an energy source. Moreover, some products from the conversion (e.g., biochar from pyrolysis) can be used as soil additive to recover nutrients and carbon to the soil. The latter can additionally act as water storage. This subject is important because there are significant quantities of CS waste from agricultural production globally, which is a potential source of revenue. Furthermore, other risks associated with cotton waste such as farm hygiene by pesticide remnants and soilborne pathogens can be addressed. Therefore, utilizing CS biomass has the potential to be a significant source of energy and an opportunity to reduce their environmental issues and financial costs [11]. This study contributes to the needed understanding of energy derived from thermal and biological conversion products of cotton stalks.

2. Cotton stalk residues for energy

Cotton stalks are a common agricultural residue with little economic value. They may be utilized without direct competition to food or feed provision. It is a renewable lignocellulosic biomass produced during cotton production. Daud et al. report values of 58.5% cellulose, 14.4% hemicellulose, and 21.4% lignin, which makes it a particularly attractive feedstock for thermochemical conversion processes [25]. Based on biomass classifications, cotton agricultural waste is a primary residue and herbaceous biomass fuel [26, 27]. Cotton crop cultivation occurs between July and February, while harvesting occurs from October to March [28]. Cotton agricultural wastes consist of the main stem, branches, bur, boll rinds, bracts, peduncle, roots, petioles, and leaf blades (**Figure 1**) left as residual biomass after harvesting the floral cotton bolls for commercial purposes, equivalent to roughly 3–5 times the weight of the produced cotton. The roots are 23.2% of the whole plant in average with the measured values ranging between 14.3 and 29.1%. However, based on observations of the amount of the soil stacked on the roots during fieldwork, it was decided in most studies to investigate the possibility of collecting only the aerial part of the residue, leaving the roots in the field. It was anticipated that the collected

Figure 1.
Cotton residual wastes after harvesting.

material would be free of soil and with less moisture content. These factors would make its storage easier and its use for energy production by thermal conversions more attractive [29].

The separated CS consists of the main stem, branches, burs, boll rinds, bracts, and peduncles [28, 30]. The stem has an outer fibrous bark weighing 20% of the weight of the stalk as well as an inner pith [11, 31]. It reaches between 1 and 1.75 m long, and the diameter above ground varies between 1 and 2.5 cm. On average depending upon species and crop conditions, 2 to 3 tons of CS are generated per each hectare of land annually; it's worth noting that the moisture content was found to drop from 50% to under 20% when the stalks were left in the field, after harvesting, for 3 weeks [28].

According to the US Department of Agriculture (USDA), the total global production of cotton was roughly 26.9 million metric tons for the reporting year of 2018 from August 1, 2017, to August 1, 2018, which has been relatively steady for the last 5 years of data collection. The three largest global producers of cotton in 2018 were India, China, and the USA. India produced 6.3 million metric tons of cotton, China produced 6.0 million tons, and the USA produced 4.5 million tons. The remaining countries produce less than 2 million tons year^{-1} [32].

To determine the total CS residue or collectable dry residue from the cotton production values, several factors are required. These are the annual production, residue to the crop factor, dry weight factor, and the availability factor [33]. The annual production is reported yearly by each respective country and collected by the USDA [32]. The residue factor is based on the ratio of the fresh weight of residue to the grain weight harvested at field moisture. It describes the relationship between crop grown for product and the residual biomass leftover after harvest. The relationship is specific to the type of crop variety [33, 34]. As mentioned previously, the residue typically weighs three to five times the harvested cotton [31]. Klass and co-workers estimate the residue factor to be 2.45 [33]. The availability factor is based on the end use of the CS residue and how much is available for collection. The availability of crop residues may be limited due to tilling some residues into the soil to reduce erosion risk; to provide structure; to preserve fertility; to use as a fertilizer, as fibrous material for various agricultural uses; or to feed to livestock [34]. Therefore, it can be best described as the sustainable removal rate of a residue [35]. Typically, in areas with low SOC, more crop residues will be tilled into the soil, while in areas with high SOC, more crops can be sustainably removed [34]. Many studies assume roughly 25% of total available agricultural residues can be recovered; however, recovery percentages may be higher or lower depending on the crop [35, 36]. It is estimated that in the USA, up to 70% of the residues are tilled back into the soil for nutrient cycling and soil health, whereas in India 15% is used for fuel, while the remainder is burned in the field [28, 30]. Klass et al. report a residue factor of 0.6 for cotton agriculture. Lastly, the dry weight factor is the amount of moisture in the freshly harvested cotton residue. Therefore, collectable dry biomass can be calculated with all of these values [33]. It is worth noting that harvesting crop residues for energy has been shown to be efficient and the energy required to collect and process residues is a small percentage of the energy content of the residue itself [37].

3. Characterization of cotton stalks for determination of energy potential

In order to get an overview of the main fuel properties of cotton stalks, proximate as well as ultimate analyses need to be performed. Schaffer et al. [24] compared data from cotton stalks to data for wheat straw and beechwood (**Table 1**). In

Properties		Unit	Basis	Biomass		
				Cotton stalks	Wheat straw	Wood (beech)
Proximate analysis	Ash	wt%	Dry	5.51	4.35	0.82
	Volatile matter	wt%	Dry	73.29	79	84
	Fixed carbon	wt%	Dry	21.20	17	15
Ultimate analysis	Carbon	wt%	Dry	47.05	47.82	48.26
	Hydrogen	wt%	Dry	5.35	5.29	5.80
	Nitrogen	wt%	Dry	0.65	0.47	0.29
	Sulfur	wt%	Dry	0.21	0.08	0.03
	Oxygen	wt%	Dry	40.77	41.59	44.80
Lower heating value (LHV)		MJ/kg	Dry	17.1	17.7	17.4

Representative values based on analyses reported in the literature [29, 38, 39]

Table 1.
Fuel properties of cotton stalks, wheat straw, and wood on a dry basis.

comparison with wood, the agricultural by-products are characterized by higher ash contents. The lower heating value (LHV) of dry cotton stalks is equivalent to poor-quality wood and varies from 16.4 to 18.26 MJ/kg [38]. Compared with wheat straw (LHV of 17.28–18.41 MJ/kg [38]), the cotton stalk can be considered as a biofuel with respect to its energy content. However, a clean and energy-efficient utilization in combustion plants is counterindicated by high contents of elements like Cl, K, and Na that decrease the ash melting point of SiO_2 and lead to fouling and corrosion in the boiler plant. Although straw and stalks are, therefore, not suitable for conventional combustion plants, low-temperature thermochemical conversion could be applied with the effect to yield biologically stable biochar containing the critical ash constituents and also plant nutrients, while the ash-free volatiles can be used in high-temperature conversion routes such as combustion in gas boilers or cofiring in pulverized coal boilers. In this respect, it is important to notice that the fixed carbon content obtained in the proximate analysis is higher for cotton stalks than wheat straw and beechwood. This observation holds true also when looking at other fuel samples available in the literature cited in **Table 1**. Furthermore, it is seen that cotton stalks possess high amounts of carbon (47.05%) and oxygen (40.77%) and its composition is relatively similar to wheat straw and wood. The presence of these elements in biomass leads to more char formation as well as to the high calorific value of the product. Therefore, because cotton stalks, wheat straw, and wood have high carbon and oxygen contents, they are suitable for energy production and could be combined with the supply of biochar.

Proviso studies have shown that raw CS provides higher combustion efficiency and longer burn time than some other agricultural residuals; furthermore, the energy needed to collect and process these residues is a small percentage of the energy contained within them [11]. To summarize, cotton stalk can be considered a typical biofuel with respect to its energy content.

4. Bioenergy conversion technologies

Bioenergy carriers are solid, liquid, or gaseous fuels which can be obtained from the available technologies. Liquid fuels are commonly used in transportation vehicles but can also be used in stationary engines especially turbines. Solid fuels are directly combusted to obtain heat, power, or combined heat and power (CHP). Gaseous fuels can be applied to the full range of end uses. As CS calorific value is equivalent to poor-quality woody biomass. A method of increasing the calorific value of the feedstock while simultaneously utilizing the residue is the technological processing through thermal and bioconversions to yield high-energy-content products which can be more easily transported and stored for use at a later time [11, 22]. CS can be converted into several useful forms of energy using different processes (conversion technologies). Bioenergy is the term used to describe energy derived from CS feedstocks. Several processing steps are required to convert raw CS into useful energy using mainly the two main process technology groups available: biochemical and thermochemical. Biochemical conversion encompasses two primary process options: anaerobic digestion (to biogas) and fermentation (to ethanol). For the thermochemical conversion routes, the four main process options presented here are pyrolysis, gasification, combustion, and hydrothermal processing (basically hydrothermal carbonization (HTC)). **Figure 2** provides a broad classification of energy conversion processes for CS.

4.1 Thermochemical conversion

Thermochemical conversion of biomass is the process of utilizing heat and, in some cases, chemical reagents, to create more energetically useful products. The output from the process is heat, gaseous, liquid, or solid fuels [40]. The four major thermal processes for converting biomass to useful energy are combustion, gasification, pyrolysis, and hydrothermal processes (see **Figure 2**). Hydrothermal processes summarize three distinct processes such as hydrothermal carbonization, liquefaction, and gasification. Hydrothermal carbonization is the process which fits best to cotton stalks and is the most developed, and therefore the focus is here on this conversion route. Pyrolysis, gasification, and combustion can be seen as state-of-the-art technologies, although not implemented in demonstration scale for cotton stalks yet. All processes can be implemented in similar plant configurations (fix bed, fluidized bed, entrained flow). Pyrolysis seems to be the most promising thermochemical conversion route due to its robustness, flexibility, and the possibility to provide a method to recover nutrients. Thus, pyrolysis is described in more details.

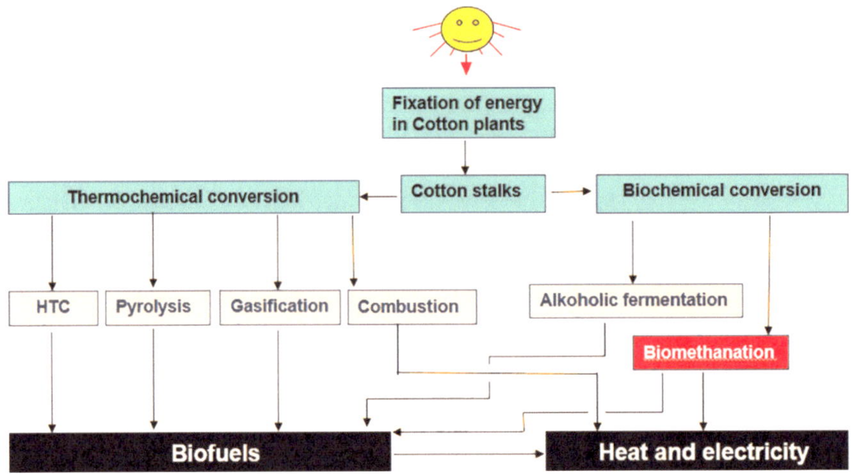

Figure 2.
Schematic diagram of the processes of energy conversion of cotton stalks.

4.1.1 Combustion

Combustion, or direct burning, of biomass consists of full oxidation of combustibles in air or oxygen-enriched air. Generally, biomass combustion produces a variety of pollutants and particulate matter (PM), as well as flue gas which requires special treatment of unburned particles. In comparison to gasification and dependent on the feedstock used for fuel, combustion can release the acid rain contributing pollutants sulfur oxides (SO_x) and nitrogen dioxide (NO_2) at roughly 40 times and 9 times, respectively [41]. Combustion of biomass with high ash content has several drawbacks in comparison with low-ash biomass. The remnant ash content is left deposited on the internal heating surfaces, which forms slags and causes fouling to the process, affecting the heating rate negatively and decreasing process efficiency [9]. The inorganic compounds in the biomass feedstock may lead to an increase in particulate matter (PM) concentrations, such as crystalline silica, which has detrimental health effects in the air [9, 42]. With consideration to the detrimental impacts of ash on combustion processes, the ash content of the CS is relatively high with 5.5 wt% db. Although straw and stalks are, therefore, not suitable for conventional combustion plants, the ash problem can be avoided by separating it into biochar through pyrolysis at low temperatures prior to combustion [9]. This can be also done by air staging in the boiler to separate the oxidation of the gases from contact to the ash. However, it has been reported in a number of studies that CS provides the highest burning efficiency and longest burn time compared to corn stover and soybean residues. The greater the density, the longer the duration of combustion. This could lead to the necessity to pelletize the feedstock for certain applications. In the study by Coates [37], it was shown that cotton plant residue could be incorporated with pecan shells to produce commercially acceptable briquettes. However, changeover of the existing factories to facilitate utilization of CS would require an initial infusion of capital. This should be compensated by lower raw material costs in a reasonable period of time.

4.1.2 Gasification

Gasification is the thermochemical conversion of biomass by partial oxidation with O_2 and the reformation by steam, carbon dioxide, or other gasification agents, producing syngas as a chemical product or fuel. The biomass is exposed to less O_2 than in combustion but more than in conditions of pyrolysis. Gasification may be allo- or autothermal; therefore, the heat required for endothermal processing is provided by ex or in situ combustion of char or gas [43]. Gasification is one of the most efficient methods for converting the chemical energy stored in biomass into heat and other useful forms of energy. Estimates of overall exergetic efficiency range from high estimates between 80.5 and 87.6% [44]. It is closely related to pyrolysis, in which both processes undergo devolatilization of biomass in the absence of O_2 or air to yield suitable products for energy without entire combustion. However, the process is optimized for maximum gas yield through oxidation and subsequent reduction [41, 44]. Gasification is processed at temperatures of typically 750–900°C for fixed and fluidized bed, 1200–1500°C for entrained flow, and up to 3000°C for plasma applications.

The products yielded by gasification include a high proportion of gases, namely, carbon monoxide (CO), carbon dioxide (CO_2), methane (CH_4), water (H_2O), hydrogen (H_2), gaseous hydrocarbons, minimal char residue, and condensed oil and tar. An oxidizing agent is added to the reaction in the form of air, O_2, or steam; and the gaseous tar or oil in the gas is condensed to acquire the desired product, producer gas. The gas may have a low energy content for autothermal operation, between 3 and 5 MJ/m^3, 10% of the heating value of natural gas; however, it is enough to power gas engines and increases the value of feedstocks that would otherwise be considered wasteful [41, 44]. For allothermal operation heating values of 12–14 MJ/m^3 are achieved. The relatively low temperature of the process leaves a char residue, which can subsequently be gasified through burning it at a high temperature, such as at 1000°C, while simultaneously injecting steam into the process. This breaks down the steam into oxygen (O_2) and hydrogen (H_2) which react with the carbon (C) from the char to create CO and H_2. By using O_2 rather than air, high-quality syngas can be produced from the CO and H_2 yield of the reaction, after impurities such as sulfur (H_2S), ammonia (NH_3), and tar have been removed. This syngas has the potential to be synthesized into methanol (CH_3OH), a high value liquid fuel, as well as other types of hydrocarbon compounds through the Fischer-Tropsch process. The efficiency of the overall process varies from 40% in simple designs to roughly 75% in processes which are well designed [41]. Allesina et al. [45] indicate that cotton residue gasification represents the basis for local circular economy models.

4.1.3 Pyrolysis

Pyrolysis is the process of thermochemical decomposition of a substance in the absence of O_2 [46]. Pyrolysis is a similar process to gasification; however, gasification controls the O_2 more precisely and generally; pyrolysis produces a significantly larger portion of biochar and is therefore sometimes called carbonization [47]. Pyrolysis is typically operated at 400–600°C. Pyrolysis produces a bio-oil liquid which can be used directly as a fuel and as a pretreatment intermediate step for converting biomass into a high-energy liquid which may be processed for power, heat, biofuels, and chemicals. Compared to the other technologies, pyrolysis is expected to offer more versatility, environmental stewardship, and higher efficiency [48]. Economically, periods of the energy crisis and fluctuating prices and availability have made biomass pyrolysis a more significant technology for development and research [49].

4.1.3.1 Pyrolysis product yields

Cotton stalk pyrolysis in a fixed-bed reactor has been studied to demonstrate products yield variation for different temperature regions [50]. They indicate that temperature increase from 650 to 800°C favored gas production, while char production decreased from 66.5 to 26.73 wt%, as the temperature increased from 250 to 650 C. This effect can be thought of as more volatile material being forced out of the char at higher temperatures, thereby reducing yield but increasing the proportion of carbon in the char. As far as the liquid fraction of the products is concerned, there is an optimum temperature at which maximum oil yield obtained (41% at ~550°C). Further temperature increases resulted in tar and liquid cracking into gases, and hence a high gas production is achieved. Similar results are also reported by [51]. The higher heating value (HHV) of pyrolysis oil is 16–23 MJ/l compared to fossil fuel which is 37 MJ/l. Pyrolysis oil has a low pH value of around 3, which must be taken into account in its handling and use. The (hydrophilic) bio-oil has water contents of typically 15–35 wt%. Typically, phase separation does occur when the water content is higher than about 30–45%.

4.1.3.2 Pyrolysis system

Pyrolysis reactors can be operated in continuous or batch mode. Typical continuous pyrolysis reactors include fluidized-bed pyrolysis, auger/screw-type pyrolysers, and rotary kilns. These reactors involve continuous input of feedstock and output of biochar, bio-oil, and syngas and often result in higher biochar yields and operational efficiencies than batch processes [52]. Compared to batch reactors, continuous reactors are more complex and expensive to design and operate and may require a reliable source of electricity [52, 53]. Therefore, continuous reactors are ideal for medium- to large-scale biochar production systems relying on centralized large quantities of feedstock. Additional information about the particularities of different pyrolysis systems can be found in the literature [48]. Nevertheless, some continuous reactor types are suitable for application in small to medium scale, too [53–56].

For the present study with cotton stalks as the feedstock, the continuously operated, indirectly heated rotary kiln reactor has been recommended according to **Figure 3**. The reasons for this decision are:

- The technology is robust and industrially proven not only for biomass but also for waste [57].

- Small- to medium-scale technology is readily available for distributed application in cotton-producing countries.

The elements chlorine and potassium, which are critical for combustion systems, remain quantitatively in the pyrolysis char fraction [55], and about 50% of the primary fuel energy can be exported with the gas and oil fraction, while less than 50% of the primary fuel energy stays in the char fraction. Thus, if the char is not further converted but returned to the soil, the problematic compounds may even have positive effects as nutrients and the related carbon will not be released as CO_2. Therefore, researchers consider the potential application of the pyrolysis char as a soil additive to increase crop yield [56, 59] or as a negative emission technology [24, 60].

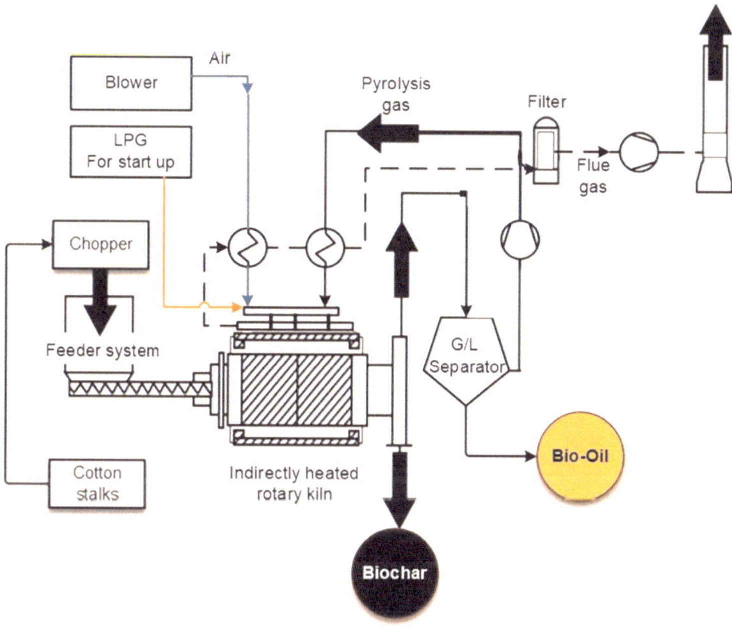

Figure 3.
Example of an indirectly heated rotary kiln pyrolysis process scheme [58].

4.1.3.3 Char utilization from cotton stalks for sustainable soil enhancement and carbon storage

Currently, the cotton stalks are often burnt on the fields causing high local pollution. However, the solid residues of the stalks remain on the field supplying nutrients. The same effects can be reached by the application of biochar from stalks. Cotton crops typically grow in hot regions on sandy soils, where biochar addition has been reported to enhance the soil fertility [59]. Mild conversion conditions below 600°C avoid ash melting and keep nutrients available for microorganisms and plants. With respect to the carbon storage effect, biochar from pyrolysis at >500°C shows sufficiently low O/C ratios to promise longevity in the soil [61]. Generally, slow pyrolysis is preferred for increased char yield [40]. The steady-state process simulation environment IPSEpro was used by Schaffer et al. [24] to assess a virtual pyrolysis conversion of cotton stalks, and they indicated that 52.8% of the carbon contained in the biomass accumulates in the biochar, whereas 38% of the input energy can be exported as heat energy at temperature levels suitable for electricity generation or industrial heat supply. The pyrolysis char shows a low molecular O/C ratio of 0.07 and an H/C ratio of 0.26. The expected half-lives of biochar in the soil are in the order of 1000 years for O/C ratios below 0.2. This makes the presented approach an interesting low-tech negative emission option. The predicted net negative emissions through stored carbon amount to 2.42 t CO_2 per hectare and year (**Figure 4**). The overall CO_2 emission avoidance effect can be increased if fossil fuel is substituted by the energy exported from the pyrolysis process.

From **Figure 5** one can see that 52.8 wt% of the total amount of carbon stored in cotton stalks is converted to char. Furthermore, the inorganic matter contained in the char, which includes important nutrients, remains in the char. The nutrients are then available for the new generation of plants if the char is used as soil additive.

The remaining part of carbon in pyrolysis gas and oil can be used for energy production as shown in the energy flow diagram in **Figure 6**. Energy streams are

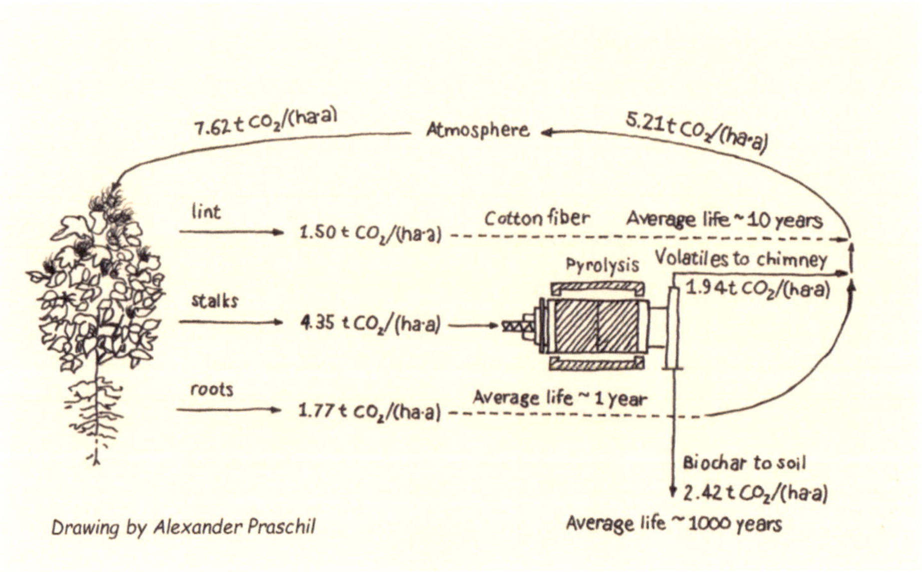

Figure 4.
Net carbon removal from the atmosphere through pyrolysis of cotton stalks and soil application of the pyrolysis char [24].

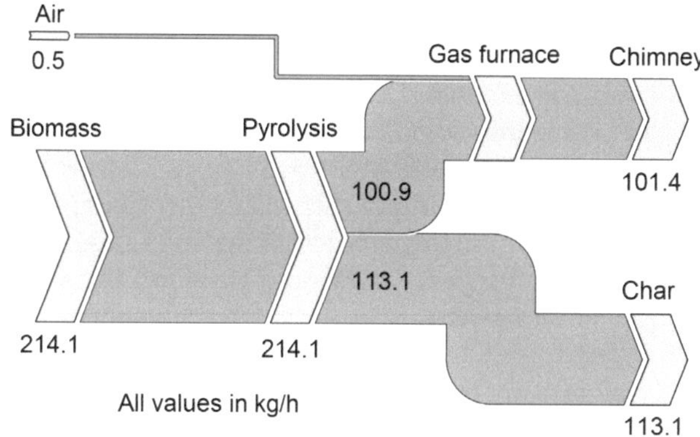

Figure 5.
Carbon mass flow diagram for an indirectly heated rotary kiln pyrolysis process without condensation of pyrolysis oil [24].

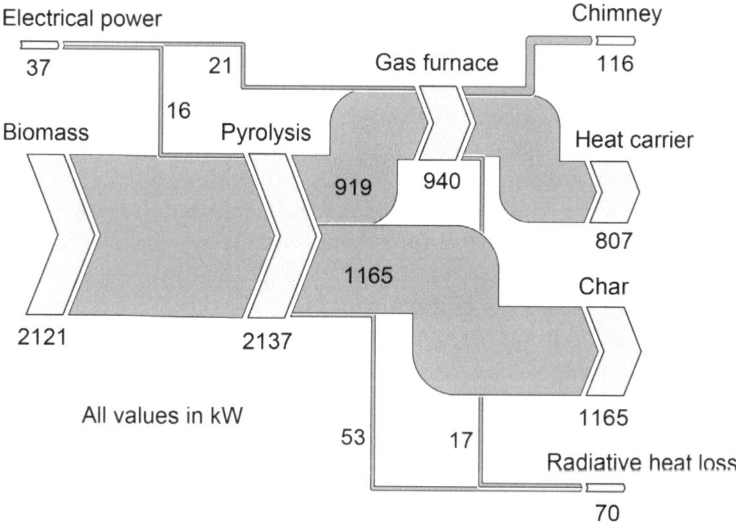

Figure 6.
Energy flow through an indirectly heated rotary kiln pyrolysis process without condensation of pyrolysis oil [24].

assessed based on the lower heating value and sensible heat with a reference temperature (sensible heat of zero) of 273.

In conclusion, the use of agricultural wastes such as cotton stalks in distributed, small- to medium-scale, energy-autonomous pyrolysis plants will allow for quasi-permanent soil storage of a part of the carbon contained in the biomass without the need for CO_2 storage sites. As a side-effect, it is expected that soil quality can be maintained and even improved by the application of biochar.

4.1.4 Hydrothermal carbonization

Nowadays hydrothermal carbonization is mentioned as a promising technology to convert biomass into a high-quality bioproduct, namely, hydrochar, as well as process water to recover nutrients (e.g., P, N, K, Si, etc.). Carbonization depletes compounds rich in oxygen and hydrogen and thereby increases the carbon content in the coal compared to the starting material. The depleted compounds are found essentially in the so-called process water and at low levels in the resulting process gas again. The product hydrochar is more hydrophobic than the source material, and the drainage is less energy intensive than the dewatering of fresh biomass. In addition, essential reactions are exothermic, and upon carbonization, after initial energy input, heat energy is released. Due to the increased carbon content of the hydrochar, the heating value increases. The hydrothermal carbonization, e.g., of CS, kills the eggs of the pink bollworm and other pathogens. There is still a need for research in the area of reduction of impurities and in the accumulation of nutrients in the coal. The distribution of nutrients between the solid, liquid, and gaseous phase can be adjusted via the process conditions (pressure, temperature, residence time, heating rate, pH, additives, catalysts, etc.). The considered process is shown in **Figure** 7.

Al Afif et al. [62] investigate the use of HTC in the production of hydrochar from CS. They concluded that hydrothermal carbonization is a promising conversion technology to provide bioenergy from CS. And there was a strong dependence between the residence time and the char quality, as the LHV of the hydrochar from CS increased with increasing residence time, whereas the total amount of hydrochar was decreased.

4.2 Biochemical conversion

Cotton stalk, as lignocellulosic biomasses, is difficult to hydrolyze due to its complex structure and a large amount of lignin present in it. Basic steps involved in bioconversion process of lignocellulosic biomass are pretreatment (physical, chemical, biological, and their combination) for cell wall destruction for biogas production, hydrolysis (acid or enzymatic) for soluble sugar release, and fermentation (bacteria or yeast) for ethanol production. Due to recalcitrant nature of lignin and its binding with holocellulose, a pretreatment step is required for fractioning of different cell wall components. Pretreatment exposes the cellulose surface for enzymatic attack and improves enzymatic digestibility and subsequent processes. Pretreatment identifies one of the major economic costs in the biochemical conversion process [63]. Generally, both process routes as discussed in the following are technically feasible, but techno-economic assessments are missing.

4.2.1 Ethanol production

The six-carbon sugars, or hexoses, glucose, galactose, and mannose, can be fermented to ethanol by many naturally occurring organisms. Baker's yeast, or *Saccharomyces cerevisiae*, has been traditionally used in the brewing industry to produce ethanol from hexoses. Recently, engineered yeasts have been reported to efficiently ferment xylose and arabinose, as well as mixtures of xylose and arabinose. In order to effectively utilize cotton stalk as a feedstock for ethanol

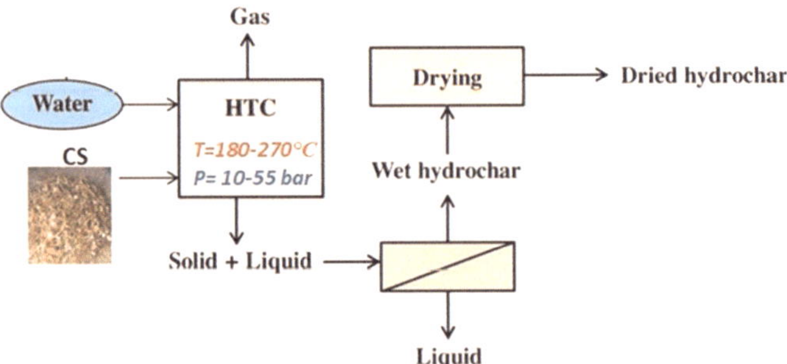

Figure 7.
System boundaries of the considered hydrothermal carbonization process.

production, optimal pretreatment is required to render the cellulose fibers more amenable to the action of hydrolytic enzymes. Generally, alkaline pretreatment is found to be more effective on agricultural residues and herbaceous crops such as cotton [64]. Christopher et al. [65] indicate that a hydrolytic efficiency of 80% was achieved for alkali-treated biomass using cellulase supplemented with beta-glucosidase and concluded that cotton stalks have great potential as a bioethanol feedstock.

4.2.2 Biogas production

Anaerobic digestion is a technology widely used for treatment of organic waste for biogas production. Biogas is a combustible gas derived from decomposing biological waste in the absence of oxygen. Biogas normally consists of 50–60% methane. It is currently captured from landfill sites, sewage treatment plants, livestock feedlots, and agricultural wastes. There were only a few studies on the subject of biogas production from cotton wastes. Isci and Demirer [23] studied the biogas production potential of cotton wastes. They indicated that cotton wastes can be digested anaerobically yielding 65–86 lN CH_4 kg^{-1} VS $(24\ days)^{-1}$. A two-stage digestion technique for biogas production from co-fermentation of organic wastes (rice, maize, cotton) was also investigated [66]. This study indicated that under anaerobic conditions from the main components in CS, the cell wall carbohydrates were well preserved, while the level of soluble carbohydrate was low. Pretreatment of lignocellulosic biomass is a necessary step to overcome the hindrance of lignin and to increase solubilization [67]. Al Afif et al. [22] investigated the anaerobic digestion of cotton stalk (CS) using organosolv plus supercritical (SC) carbon dioxide pretreatment of cotton stalks for methane production. Results indicated that supercritical carbon dioxide pretreatment of CS is a potential option for improving the energy output, as the pretreatment of CS samples with organosolv plus SC-CO_2 increased the methane yield up to 20% compared with the untreated samples. The highest methane yield of 177 lN kg^{-1} VS was achieved by pretreatment with organosolv plus SC-CO_2 at 100 bars and 180°C for 140 minutes. It is worth noting that the quality of biogas was good and increased with pretreatment from 50 to 60% CH_4. To summarize, cotton stalks can be digested anaerobically and is a good source of biogas; nevertheless, pretreatment of cotton stalks is a necessary step to increase solubilization hence the methane production.

4.3 Future perspectives

This study contributes to enhancing our understanding of the feasibility of bioenergy recovery from cotton stalks. The findings have the potential to lead to a sustainable solution for the treatment of cotton stalks.

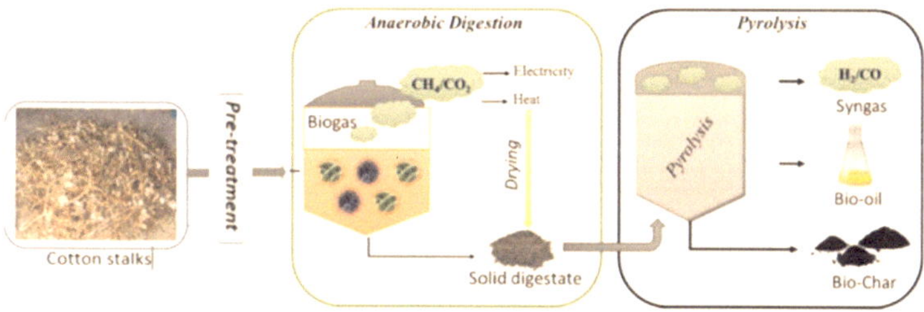

Figure 8.
The system boundary of coupling anaerobic digestion and pyrolysis process.

However, for higher bioenergy recovery, a study of the techno-economic feasibility of the integrated processes of anaerobic digestion and pyrolysis is recommended (see **Figure 8**).

5. Conclusions

It has been shown in this study that:

- CS is an agricultural residue with low economic value, and there is no direct competition to food or feed provision.

- CS contains lignin and carbohydrates, like cellulose and hemicelluloses, which can be converted into a variety of usable forms of energy.

- CS is more appropriate for the production of energy pellets due to its woody structure; however, due to the ash content of the CS which is relatively high with 5.5 wt% db, the ash problem can be avoided by separating it into biochar through pyrolysis at low temperatures prior to combustion.

- The use of pyrolysis and hydrothermal processes for CS treatment would result in the conversion of the major amount of carbon to char, which would mean a significant decrease in CO_2 release, compared to the state-of-the-art treatment paths. Also using biochar in the soil will reduce the need for mineral fertilizer since nutrients return to the soil with the char.

- CS can be digested anaerobically and is a good source of biogas or fermented to produce ethanol. However, pretreatment of cotton stalks is a necessary step to increase solubilization hence the methane and ethanol production.

The findings have the potential to lead to a sustainable solution for the treatment of cotton stalks. However, for higher bioenergy recovery more studies are needed to prove the effectiveness of cotton waste utilization.

Author details

Rafat Al Afif*, Christoph Pfeifer and Tobias Pröll
Institute of Chemical and Energy Engineering, University of Natural Resources and Applied Life Sciences, Vienna, Austria

*Address all correspondence to: rafat.alafif@boku.ac.at

References

[1] International Energy Agency (IEA). Key world Energy Statistics. Paris: International Energy Agency; 2016. Available from: https://www.iea.org/publications/freepublications/publication/KeyWorld2017.pdf

[2] Budinis S, Mac Dowell N, Krevor S, Dixon T, Kemper J, Hawkes A. Can carbon capture and storage unlock 'unburnable carbon'? Energy Procedia. 2017;**114**:7504-7515. ISSN 1876-6102

[3] BP. BP Statistical Review of World Energy June 2017. 2017. Available from: https://www.bp.com/content/dam/bp/en/corporate/pdf/energy-economics/statistical-review-2017/bp-statistical-review-of-world-energy-2017-full-report.pdf

[4] Tripathi M, Sahu JN, Ganesan P. Effect of process parameters on production of biochar from biomass waste through pyrolysis: A review. Renewable and Sustainable Energy Reviews. 2016;**55**:467-481

[5] Renewable Energy Policy Network for the 21st Century (REN21). Renewables 2018: Global Status Report. Paris: REN21 Secretariat; 2018. Available from: http://www.ren21.net/wp-content/uploads/2018/06/17-8652_GSR2018_FullReport_web_final_pdf

[6] Edenhofer O, Pichs R, Madruga Sokona Y, Seyboth K, Matschoss P, Kadner S, et al. editors. Special report on renewable energy sources and climate change mitigation. Intergovernmental Panel on Climate Change. ISBN: 978-92-9169-131-9.2012

[7] Gasparatos A, Doll CNH, Esteban M, Ahmed A, Olang TA. Renewable energy and biodiversity: Implications for transitioning to a Green Economy. Renewable and Sustainable Energy Reviews, Volume. 2017;**70**(2017): 161-184. ISSN 1364-0321

[8] Xu J, Li M. Innovative technological paradigm-based approach towards biofuel feedstock. Energy Conversion and Management. 2017;**141**:48-62. ISSN 0196-8904

[9] Dunnigan L, Ashman PJ, Zhang X, Kwong CW. Production of biochar from rice husk: Particulate emissions from the combustion of raw pyrolysis volatiles. Journal of Cleaner Production. 2018;**172**: 1639-1645. ISSN 0959-6526

[10] Obour PB, Jensen JL, Lamandé M, Watts CW, Munkholm LJ. Soil organic matter widens the range of water contents for tillage. Soil and Tillage Research. 2018;**182**:57-65. ISSN 0167-1987

[11] Hamawand I, Sandell G, Pittaway P, Chakrabarty S, Yusaf T, Chen G, et al. Bioenergy from cotton industry wastes: A review and potential. Renewable and Sustainable Energy Reviews. 2016;**66**: 435-448. ISSN 1364-0321

[12] Dunnigan L, Morton BJ, Ashman PJ, Zhang X, Kwong CW. Emission characteristics of a pyrolysis-combustion system for the co-production of biochar and bioenergy from agricultural wastes. Waste Management. 2018;**77**:59-66. ISSN 0956-053X

[13] CABI. Invasive Species Compendium: *Pectinophora gossypiella* (Pink Bollworm). 2018. Available from: https://www.cabi.org/isc/datasheet/39417#7BD9E856-BD8A-4AA5-9060-4E6504EC8EEC

[14] Lykouressis D, Perdikis D, Samartzis D, Fantinou A, Toutouzas S. Management of the pink bollworm *Pectinophora gossypiella* (Saunders) (Lepidoptera: Gelechiidae) by mating disruption in cotton fields. Crop Protection. 2005;**24**(2):177-183. ISSN 0261-2194

[15] El Saeidy E. Renewable Energy in Agriculture in Egypt: Technological Fundamentals of Briquetting Cotton Stalks as a Biofuel. Master [thesis]. Humboldt University of Berlin; 2004. Available from: https://edoc.hu-berlin.de/bitstream/handle/18452/15724/El-Saeidy.pdf?sequence=1

[16] Ellsworth P, Moor L, Allen C, Beasley B, Henneberry T, Carter F. Pink Bollworm Management. 2003. Available from: http://ag.arizona.edu/crops/cotton/insects/pbw/NCCPBWnewsNo2.pdf

[17] Blake C.. Western Farm Press: Pink Bollworm Eradication in Cotton in 2017? 2014. Available from: https://www.westernfarmpress.com/cotton/pink-bollworm-eradication-cotton-2017

[18] Heilmann S, Davis H, Jader L, Lefebvre P, Sadowsky M, Schendel F, et al. Hydrothermal carbonization of microalgae. Biomass & Bioenergy. 2010; **34**:875-882

[19] Bridgwater A. Biomass fast pyrolysis. Thermal Science. 2004;**8**: 21-49

[20] Algin H, Turgot P. Cotton and limestone powder wastes as brick material. Construction and Building Materials. 2008;**22**:1074-1080

[21] Zhang Y. Hydrothermal liquefaction to convert biomass into crude oil. In: Blaschek HP, editor. Biofuels from Agriculture Wastes Byproduct. Oxford: Blackwell Publishing; 2010. pp. 201-232

[22] Al Afif R, Wendland M, Krapf LC, Amon T, Pfeifer C. Organosolv plus supercritival carbon dioxide pre-treatment of cotton stalks for methane production. In: Proceedings of the 10th International Conference on Sustainable Energy and Environmental Protection, Bioenergy and Biofuels. University of Maribor Press; 2017. pp. 22-33. ISBN: 978-961-286-048-6

[23] Isci A, Demirer G. Biogas production potential from cotton wastes. Renew Energy. 2007;**32**(5): 750-757

[24] Schaffer S, Pröll T, Al Afif R, Pfeifer C. A mass- and energy balance-based process modelling study for the pyrolysis of cotton stalks with char utilization for sustainable soil enhancement and carbon storage. Biomass and Bioenergy. 2019;**120**: 281-290

[25] Daud Z, Hatta MZM, Kassim ASM, Awang H, Aripin AM. Analysis the chemical composition and fiber morphology structure of corn stalk. Australian Journal of Basic and Applied Sciences. 2013;7(9):401-405. ISSN 1991-8178

[26] European Commission. Intelligent Energy Europe. Biomass Energy Centre. Solid Biofuels. Fuel Specifications and Classes. Graded Non-woody Pellets: Solid Biofuels. Fuel Specifications and Classes. Graded Non-woody Pellets. 2014. Available from: https://ec.europa.eu/energy/intelligent/projects/sites/iee projects/files/projects/documents/whs_summary_of_woodfuel_standards_en.pdf

[27] Khan AA, de Jong W, Jansens PJ, Spliethoff H. Biomass combustion in fluidized bed boilers: Potential problems and remedies. Fuel Processing Technology. 2009;**90**(1):21-50. ISSN 0378-3820

[28] Patil PG, Gurjar RM, Shaikh AJ, Balasubramanya RH, Paralikar KM, Varadarajan PV. Cotton Plant Stalk—An Alternative Raw Material to Board Industry. 2007. Available from: https://wcrc.confex.com/wcrc/2007/techprogram/P1506.HTM.

[29] Wang X, Qin G, Chen M, Wang J. Microwave-assisted pyrolysis of cotton stalk with additives. BioResources. 2016; **11**(3):6125-6136

[30] He Z, Zhang H, Tewolde H, Shankle M. Chemical characterization of cotton plant parts for multiple uses. Agricultural & Environmental Letters. 2017;**2**: 110044. DOI: 10.2134/ael2016.11.0044

[31] Reddy N, Yang Y. Innovative Biofibers from Renewable Resources. Heidelberg, New York, Dordrecht, London: Springer; 2015. ISBN 978-3-662-45136-6

[32] United States Department of Agriculture (USDA). Cotton: World Markets and Trade 2018. Available from: https://apps.fas.usda.gov/psdonline/circulars/cotton.pdf

[33] Klass DL. Chapter 5—Waste biomass resource abundance, energy potential, and availability. Biomass for Renewable Energy, Fuels, and Chemicals, Academic Press, Pages. 1998: 137-158. ISBN 9780124109506

[34] European Commission (EC). 2016. Maximising the yield of biomass from residues of agricultural crops and biomass from forestry. Available from: https://ec.europa.eu/energy/sites/ener/files/documents/Ecofys%20%20Final%20report_%20EC_max%20yield%20biomass%20residues%2020151214.pdf

[35] Scarlat N, Martinov M, Jean-François Dallemand JF. Assessment of the availability of agricultural crop residues in the European Union: Potential and limitations for bioenergy use. Waste Management. 2010;**30**(10): 1889-1897. ISSN 0956-053X

[36] Hoogwijk M, Faaij A, van den Broek R, Berndes G, Gielen D, Turkenburg W. Exploration of the ranges of the global potential of biomass for energy. Biomass and Bioenergy. 2003;**25**(2): 119-133. DOI: 10.1016/S0961-9534(02)00191-5

[37] Coates WE. Using cotton plant residue to produce briquettes. Biomass and Bioenergy. 2000;**18**:201-208

[38] ECN. Phyllis2—Database for Biomass and Waste. 2019. Available from: www.ecn.nl/phyllis2

[39] Ahmed FA, Abdel-Moein NM, Mohamed AS, Ahmed SE. Biochemical studies on some cotton by products Part I—Chemical constituents and cellulose extraction of Egyptian cotton stalks. Journal of American Science. 2010;**6**: 1306-1313

[40] Bridgwater AV. Pyrolysis—State of the art. In: Bridgwater AV, editor. Success & Visions for Bioenergy: Thermal Processing of Biomass for Bioenergy, Biofuels and Bioproducts. Newbury, Berks, UK: CPL Press; 2007

[41] Boyle G. Renewable Energy: Power for a Sustainable Future. Oxford, England: Oxford University Press in association with the Open University; 2012

[42] Johansson LS, Tullin C, Leckner B, Sjövall P. Particle emissions from biomass combustion in small combustors. Biomass and Bioenergy. 2003;**25**(4):435-446. ISSN 0961-9534

[43] Salatino P, Solimene R. Mixing and segregation in fluidized bed thermochemical conversion of biomass. Powder Technology. 2017;**316**:29-40. ISSN 0032-5910

[44] Puig-Arnavat M, Bruno JC, Coronas A. Review and analysis of biomass gasification models. Renewable and Sustainable Energy Reviews. 2010; **14**(9):2841-2851. ISSN 1364-0321

[45] Allesina G, Pedrazzi S, Allegretti F, Morselli N, Puglia M, Santunione G, et al. 2018. Gasification of cotton crop residues for combined power and biochar production in Mozambique. Applied Thermal Engineering. 2018; **139**(5):387-394

[46] Kung CC, Zhang N. Renewable energy from pyrolysis using crops and

agricultural residuals: An economic and environmental evaluation. Energy. 2015; **90**(Part 2):1532-1544. ISSN 0360-5442

[47] Mohan D, Pittman C, Steele P. Pyrolysis of wood/biomass for bio-oil: A critical review. Energy. 2006;**20**:848. DOI: 10.1021/ef0502397

[48] Bridgwater AV. Review of fast pyrolysis of biomass and product upgrading. Biomass and Bioenergy. 2012;**38**:68-94. ISSN 0961-9534

[49] Qi Z, Jie C, Teijun W, Ying X. Review of biomass pyrolysis oil properties and upgrading research. Energy Conversion and Management. 2007;**48**:87-92

[50] Chen Y, Yang H, Wang X, Zhang S, Chen H. Biomass-based pyrolytic polygeneration system on cotton stalk pyrolysis: Influence of temperature. Bioresource Technology. 2012;**107**: 411-418. ISSN 0960-8524

[51] Kantarelis E, Zabaniotou A. Valorization of cotton stalks by fast pyrolysis and fixed bed air gasification for syngas production as precursor of second generation biofuels and sustainable agriculture. Bioresources Technology. 2009;**100**:942-947

[52] Brown R. Biochar production technology. In: Lehmann J, Joseph S, editors. Biochar for Environmental Management: Science and Technology. London: Earthscan; 2009. pp. 127-139

[53] Duku MH, Gu S, Hagan E. Biochar production potential in Ghana—A review. Renewable and Sustainable Energy Reviews. 2011;**15**:3539-3551

[54] Halwachs M, Kampichler G, Kern S, Pröll T, Hofbauer H. Rotary kiln pyrolysis—A comprehensive approach of operating a 3 MW pilot plant over a period of two years. In: Bridgwater AV, editor. Proceedings of the Bioten Conference on Biomass, Bioenergy and Biofuels. Newbury: CPL Press; 2010. p. 2011

[55] Kern S, Halwachs M, Kampichler G, Pfeifer C, Pröll T, Hofbauer H. Rotary kiln pyrolysis of straw and fermentation residues in a 3 MW pilot plant—Influence of pyrolysis temperature on pyrolysis product performance. Journal of Analytical and Applied Pyrolysis. 2012;**97**:1-10

[56] Pandit NR, Mulder J, Hale SE, Schmidt HP, Cornelissen G, Cowie A. Biochar from Kon-Tiki flame curtain and other kilns: Effects of nutrient enrichment and kiln type on crop yield and soil chemistry. PLoS One. 2017;**12**: e0176378. DOI: 10.1371/journal.pone.0176378

[57] Meier M, Schmid K, Heger S. Pyrolyseverfahren in Burgau—eine Betrachtung aus Sicht der Überwachungsbehörde. In: Thomé-Kozmiensky KJ, Beckmann M, editors. Energie aus Abfall (in German). Vol. 11. Nietwerder, Germany: TK Verlag; 2014. pp. 780-795

[58] Pröll T, Al Afif R, Schaffer S, Pfeifer C. Reduced local emissions and long-term carbon storage through pyrolysis of agricultural waste and application of pyrolysis char for soil improvement. Energy Procedia. 2017; **114**:6057-6066

[59] Schmidt HP, Pandit BH, Cornelissen G, Kammann CI. Biochar-based fertilization with liquid nutrient enrichment: 21 Field trials covering 13 crop species in Nepal. Land Degradation & Development. 2017;**28**:2324-2342

[60] Pröll T, Zerobin F. Biomass-based negative emission technology options with combined heat and power generation. Mitigation and Adaptation Strategies for Global Change. 2019. DOI: 10.1007/s11027-019-9841-4

[61] Spokas KA. Review of the stability of biochar in soils: Predictability of O:C

molar ratios. Carbon Management. 2010;**1**:289-303

[62] Al Afif R, Pfeifer C, Ramadan M. An experimental study on hydrothermal carbonization of cotton stalks. In: Proceedings of the 11th International Conference on Sustainable Energy & Environmental Protection, Bioenergy and Biofuels. Glasgow, UK; 2018

[63] Alvira P, Tomás-Pejó E, Ballesteros MM, Negro J. Pretreatment technologies for an efficient bioethanol production process based on enzymatic hydrolysis: A review. Bioresource Technology. 2010;**101**:4851-4861

[64] Silverstein RA, Chen Y, Sharma-Shivappa RR, Boyette MD, Osborne J. A comparison of chemical pretreatment methods for improving saccharification of cotton stalks. Bioresource Technology. 2007;**98**:3000-3011

[65] Christopher M, Mathew AK, Kumar KM, Pandey A, Sukumaran R. A biorefinery-based approach for the production of ethanol from enzymatically hydrolysed cotton stalks. Bioresource Technology. 2017;**242**: 178-183

[66] El-Shinnawi MM, El-Houssieni M, Aboel-Naga SA, Fahmy S. Chemical changes during a two-stage digestion technique for biogas production from combinations of organic wastes. Resour Conserv Recycling. 1990;**3**:217-230

[67] Zheng Y, Shi J, Cheng Y. Principles and development of lignocellulosic biomass pretreatment for biofuels. Advances in Bioenergy. 2017;**2**:1-68

Chapter 5

Analyzing Chemical and Physical Variations in Selected Cotton Wires at Ambient Temperatures and Conditions

Clara Silvestre de Souza and José U.L. Mendes

Abstract

Cotton, a hydrophilic textile fiber, has unstable characteristics and, for this reason, it varies its properties according to the environmental changes. Moisture and temperature are the two most important factors that lead a cotton spinning sector and influence its quality. Those two properties can change the entire spinning process. Understanding this, moisture and temperature must be kept under control during the spinning process; once the environment is hot and dry, the cotton yarns absorb moisture and lose the minimal consistency. According to this information, this chapter was developed testing four types of cotton yarns, one kind of cotton from Brazil and the others from Egypt. The yarns were exposed to different temperatures and moisture in five different tests, and in each test, six samples were examined through physical and mechanical tests: resistance, strength, tenacity, yarn's hairiness, yarn's evenness, and yarn's twisting. All the analyses were accomplished at Laboratório de Mecânica dos Fluidos and at COATS Corrente S.A., where it was possible to use the equipment which were fundamental to development of this chapter, such as the STATIMAT ME, which measures strength, tenacity; Zweigler G566, which measure hairiness in the yarn; a skein machine; and a twisting machine. The analyses revealed alterations in the yarn's characteristics in a direct way; for example, as moisture and temperature were increased, the yarn's strength, tenacity, and hairiness were increased as well. Having the results of all analyses, it is possible to say that with a relatively low temperature and high humidity, cotton yarns have the best performance.

Keywords: temperature, moisture, cotton yarn, yarn twisting, yarn hairiness

1. Introduction

The cotton fiber is the basic raw material for cotton spinning mills in cotton wires production. This fiber is the largest input material for cotton wire production used in clothing due to mainly comfort characteristics it contains. Cotton, being a natural fiber, possesses several irregularities in its physical characteristics and plenty of impurities during harvesting and ginning process. From the cotton harvesting to the yarn, some processes are necessary, and they are known as the cotton mill process.

IntechOpen

The lack of quality and the incidence of imperfections on cotton wires may result in major financial losses for the company that produces them and for the client who buys them. For this reason, the objective of a successful spinning is to produce high quality cotton yarns, economically. It is only possible by the oriented and controlled utilization of the raw material in the spinning process.

The humidity and the temperature are very important factors in a textile industry. Given that the cotton is a hygroscopic fiber, it is capable of absorbing a great amount of water, and it is necessary to keep it conditioning in an environment with relatively high humidity. To avoid fires or rupture accidents, the air temperature should be relatively medium, around 20–25°C. The humidity measured inside a cotton mill is, indeed, the relative humidity of the air, which means relative humidity is the water vapor ratio present in the air and the percent (%) of the maximum possible water vapor quantity that is currently in the air. As the maximum humidity is temperature dependent, relative humidity changes with the temperature even when absolute humidity remains constant.

With the objective of analyzing cotton wires, derived from the ring spinning, and the effect caused by the temperature and the humidity in its manufacturing process, this write-up was conducted in partnership with Universidade Federal do Rio Grande do Norte and COATS Corrente S.A.. The analyses carried had the aim of testing the different behaviors presented by cotton yarns when exposed to distinct values of temperature and humidity. The characteristics analyzed were count, nepiness, twist threads, and tensile strength. The experiments were performed in the COATS Corrente S.A. Laboratory of Quality Control. The tests were carried using three types of cotton material: Giza 88 carded, Giza 86 (carding and drawing), and Meridional carded. For each type of cotton, six bobbins were used. Daily analyses were conducted, and there were changes in the temperature and the humidity for each test day. After the analyzing the material and observing the results, it has been found that the carding and drawing wires submitted to different values of temperature and humidity presented strength and nepiness values directly proportional to the temperature increase.

Hence, this work was aimed to analyze physical and mechanical properties of cotton yarns, deriving from ring spinning, suffering temperature and humidity variations.

2. Bibliographic review

Cotton consists in a fibrous material surrounding the seeds of cotton, belonging to the mallow family, *Gossypium* spp. It is used in textile industry and is a plant of hot weather that cannot handle cold. According to the older documents, it is originated in India, expanded through Iran and Western Asia, in a westerly and northerly direction.

Cotton can be classified in various ways, including length, type of cotton, uniformity of length, index or contents of short fibers (%), fineness, resiliency, strength, elasticity, flexibility, reliability, balance, maturity, humidity and regain, color, shine, and light reflectance.

With respect to mechanical characteristics of the cotton fibers, they are probably the most important properties, contributing for fibers behavior during the process, and its properties on final product. Due to their magnitude, the most important mechanical properties of the fibers are the tensile strength properties, its behavior under the application of forces, and deformities along the axis of the fiber. One of these, the easier to study, experimentally, is the extension (elongation), and finally rupture, under a load that increases in ascending order.

The tensile strength is a measurement of the force applied in order to break the yarn or the fiber. For a single yarn, the strength is measured according to the breaking load, and when multiple twisted yarns are tested, the breaking load is also called tenacity. When the elongation is measured before breaking the yarn, the most important point to verify is the tension that every yarn may sustain, as it may be observed in the results. Another mechanical property tested was the rupture point, the necessary energy to burst the yarn.

Yarns are textile material constituted by natural or manufactured fibers, presenting fineness and high ratio of length, formed through various spinning operations. They are characterized by their regularity, diameter, and weight; the last two characteristics determine the count of the wire. Overall, the yarn can be defined as a grouping of linear fibers or filaments, which form a continuous line with textile characteristics. These textile characteristics include high tensile strength and high flexibility.

The staple fibers or filament wires, flat and textured, could be twisted with the purpose of increasing water resistance. Made from a single yarn, or inside a single yarn, a variety of effects in the wire may be created, **Figure 1** [1].

The cotton yarns used in this study were derived from the ring spinning. The bobbins containing the wires were donated by COATS Corrente S.A., which manufactures, mainly, sewing and embroidery threads. The single yarns manufactured and used in the analyses were three types: Giza 88 combed (200 dtex), Giza 86 carded and combed (220 and 222 dtex), and Meridional carded (660 dtex). For each analysis, six bobbins had been used for each type of cotton.

The ring spinning process occurs when the cards output the cards output fuses fibers and the input on a runner, where they are duplicated herewith with other fuses and then, join together again to form a fuse, this function is used to reduce mass variation per length. After that the fuses go to the roving machine in order to twist the fuse into a wick, and then wind on bobbins throughout a ring in a spinning machine. In the final part, they go from the bobbins to the cones on the cone winder. The yarn made by this method is named carded yarn. With conventional spinning, it is also possible to manufacture a combed yarn. The difference during the process is an addition of two more machines after carding, which are the Assembling Winder and the Comber, whose main function is to extract short fibers resulting in the production of high quality yarns, less nepiness, and higher strength, beside the production on very fine yarns [2].

Aiming to avoid yarns break during the spinning process, a properly humidified area and a significant cooling are necessary. The control of the room humidity ensures an increase in production due to a superior strength and flexibility of the

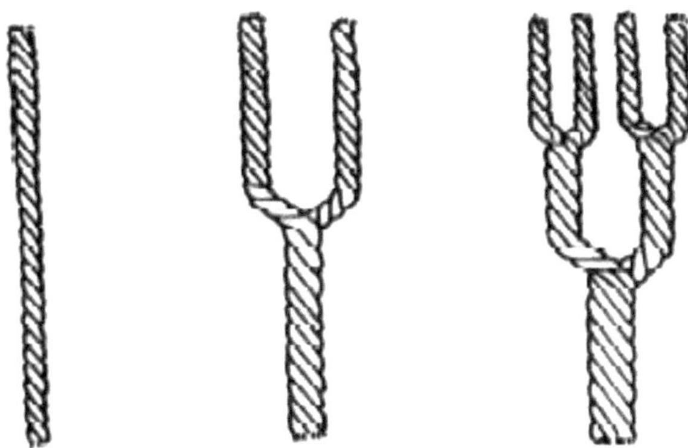

Figure 1.
Single twisted yarn; multiple twisted yarn; multiple twisted yarn twice twisted [1].

textile fibers when processed between 65 and 70% relative humidity. Furthermore, these conditions guarantee the reduction of air dust. Most textile fibers, especially natural fibers, are hygroscopic, that is to say, they are capable of absorbing or releasing humidity. When there is the excess or lack of humidity in the environment, the fiber's physical characteristics like weight, strength, etc., are altered. For example, linen and cotton show a considerable strength increase when their humidity increases. The quantity of humidity in a fiber sample may be described as Regain or recovery, or even in terms of humidity content [3, 4].

Cotton fiber Regain may change physical properties and interfere with the capacity of spin of the fiber, hence altering the results. In addition, it is known that humidity distributions along the yarn are not homogenous; therefore, humidity changes, imperfections, and defect levels shall be expected. Preconditioning in a dry atmosphere for several days and a subsequent conditioning for, at least 24 h, are some cares that must be observed [5]. Thereby, any influence resulting from thermo hygrometers conditions is completely eliminated. Therefore, textile fibers hygroscopicity is a remarkable property because of its effect on textile articles properties. Leonardo Da Vinci (1452–1519) once mentioned the property hygroscopicity in his notebooks, imagining a scale in which one plate e put cotton an in the other one, an equal mass of wax.

3. Methodology

For the carried out experiments, six bobbins (100% cotton) were donated by COATS Corrente from four types of cotton, which were Giza 88 combed 200 dtex, Giza 86 carded 222 dtex, Giza 86 combed 220 dtex, and cotton Meridional combed 660 dtex. The cotton characteristics may be checked in **Table 1**.

To get the analysis conducted, a refrigerator Consul, 120 l capacity, has been used. To take measures, a Thermo Hygrometer Oregon Scientific, temperature variation at −20 to 60°C and relative humidity 25 to 90%, has been chosen.

By choosing the refrigerator as a climatic chamber, some difficulties appeared; for example, the fact the equipment loses relative humidity to the same extent the temperature decreases. Also the abrupt reduction of the temperature compromised the research. Aiming to find the righteous temperature and humidity required by the chapter, there has been a lot of research conducted on the equipment intending to adapt the refrigerator into a climatic chamber.

After the equipment acquisition, first, analyses were made to test its capacity. The refrigerator has a thermostat for three different temperatures: MIN, MAX, and MED, as shown in **Table 2**. The refrigerator analyses were conducted in Laboratório de Mecânica dos Fluídos at Núcleo de Tecnologia da Universidade Federal do Rio Grande

COTTOM	COLOR	LENGTH	MICRONAIRE	WEIGHT
GIZA 88	SLIGHTLY YELLOW	38 – 40 mm	4.3	110 g
GIZA 86	WHITE	34 – 36 mm	4.75	110 g
MERIDIONAL	SLIGHTLY YELLOW	30 – 32 mm	3.8	110 g

Source: Author

Table 1.
Types of cotton used in the analysis and their characteristics.

THERMOSTAT POSITION	TEMPERATURE
MAX	± 5°C
MED	± 3°C
MIN	0°C

Table 2.
Temperature corresponding to the thermostat position.

do Norte, and, the yarn analyses, in Laboratório de Controle de Qualidade at COATS Corrente, in November and December, 2010, and in February and April, 2011.

After an overall evaluation of their results, it that the temperature which is best suited to analyses is the temperature reached with the thermostat on MAX. This was owed to the coordination with Brazilian technical standard NBR8428, which sets out that the ambient temperature must be around 20°C and relative humidity 65%.

With temperature variations analyses accomplished, there was a need to keep constant humidity inside the refrigerator. It was complicated, since the refrigeration equipment works taking off the humidity from the room where it is planning to refrigerate, leaving the climatic chamber dry. As there was an absence in this respect, it was necessary to supply it. The first analysis was carried in order to measure the decrease in humidity inside the refrigerator.

It could be verified, as was expected, that the temperature has dropped quickly inside the equipment, reaching approximately 0°C within 2 h.

Subsequently, a nebulizer has been inserted inside the refrigerator with an aim to increase the humidity inside the refrigerator. Nebulizer has also the task of increasing the temperature and stabilizing it.

Utilizing the nebulizer, relative humidity remained stable between 54 and 57%; however, the temperature continued to fall. Subsequently, temperature stabilization was analyzed. Humidity had a slight stabilization although it was considerable as compared with the previous day. To achieve this goal, a more complex test has been done. With the nebulizer turned on, the refrigerator had a small opening near the door, 8 cm, which permitted the equipment to exchange heat with the environment.

With the opening of the door, the temperature increases immediately reaching 0°C; however, humidity remains the same. About 2 h and 17 min of experiment later, it can be verified that temperature ranged and after stabilized, with the door opened, the humidity stabilized but decreased again.

Previous reviews have proved that nebulizer can sustain a stable humidity although it was not enough. It was also substantiated that temperature stabilizes with the door opened, exchanging heat with the external environment. Therefore, it was decided that the standard would be the use of two beakers (500 and 1000 ml) and, aiming to increase the temperature inside the refrigerator from 20 to 22°C, the door would be opened from 8 to 16 cm.

After the temperature and humidity analysis were made inside the climatic chamber, the equipment has been taken to COATS Corrente in Extremoz, Rio Grande do Norte. In the Laboratório de Controle da Qualidade da empresa, it was possible to conduct other analyses regarding tensile strength, nepiness, count, and torsion on cotton yarns.

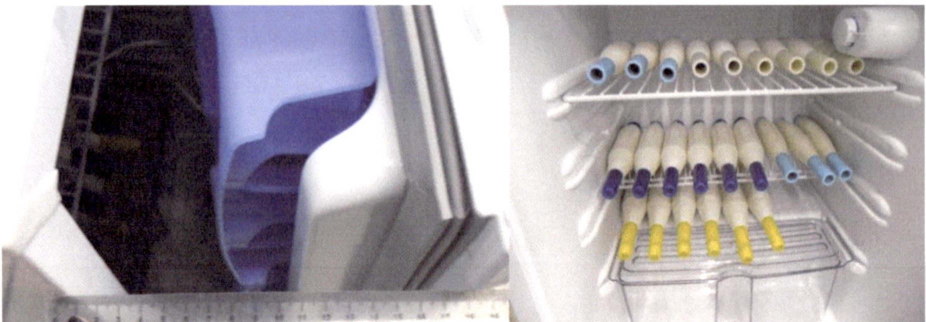

Figure 2.
Refrigerator with the door opened in 8 cm and bobbins inside the refrigerator. Source: Author

Analyses and tests were carried in COATS Corrente with a time interval of 24 h, under the NBR 8428 standard specifications. After the confinement inside the refrigerator, **Figure 2**, bobbins follow to: count, torsion, tensile strength, and nepiness. Initially, with temperature 27°C, 24 bobbins were taken directly from the spinning mill to the Laboratório de Controle da Qualidade to conduct the analyses, in an atmosphere of 27°C and relative humidity 45%. Next, bobbins were placed for 24 h in a 27°C temperature and relative humidity 33% inside the disconnected chamber.

Regarding the equipment's used for testing the refrigerated bobbins, the STATIMAT ME was used to conduct tensile strength tests on wires and for nepiness, Zweigle G566 was used. The Hairiness Tester Zweigle evaluates the amount of irregularities per millimeter on the yarn.

The count tests were carried using a winder and an electronic scale. The winder, as main purpose, creates skeins from yarns wrapped on bobbins with a predetermined length. The COATS tests, the Giza 86 and 88 winds had 100 m long and cotton, 50 m long, because of its count, thicker.

During the test, a certain length was placed between the two claws of the torsion machine and then, one of it twists and the other gnarled the yarn.

The tests conducted in COATS were carried using company's own standards, i.e., COATS quality standard.

4. Results

The main objective of the tests analyzed in this chapter was to observe physical characteristic behaviors of cotton yarns under temperature and humidity alterations. To this end, tests were carried on the wires, varying on the fluctuation of temperature and the relative humidity of the air, **Table 3**.

4.1 Tensile strength

Tensile strength tests were divided in two: tenacity is the energy required to fracture or break a part of a sample and describes its ability to absorb energy before rupture. Elongation at fracture is also a quite important factor for the wire quality since it qualifics yarns as resistant or strong, and rupture force which is characterized by the required quantity of energy to break the wire. All tests occurred in the same equipment.

To analyze tenacity, a behavior already expected was shown by the bobbins but with higher values on the last day of tests, which means, lowest relative humidity of the room, due to air conditioning.

ANALYSES	TEMPERATURE	HUMIDITY
1st Test	27°C	45%
2dd Test	27°C	33%
3 rd Test	17,4°C	32%
4 th Test	19,5°C	31%

Source: Author

Table 3.
The values of temperature and humidity for each experiment day.

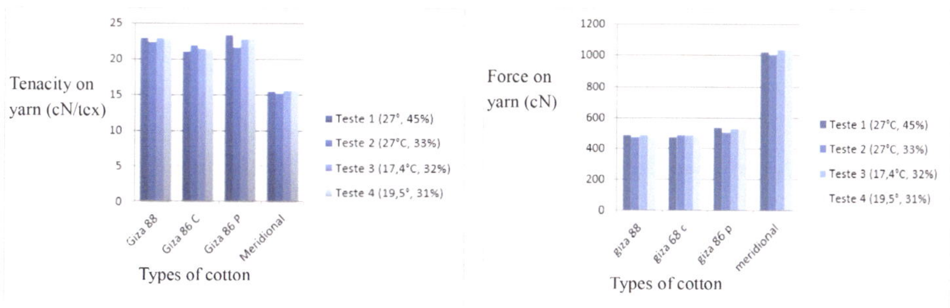

Figure 3.
Graphs related to (a) tenacity and (b) rupture force. Elongation graphs is not presented because of a similar behavior to tensile strength which has occurred. A similar performance occurred with rupture force and work. Source: Author

On the other days, the values were lower due to variation in temperature and humidity. The test proved that Giza cotton has a better quality than Meridional cotton [6].

Elongation at fracture analysis has shown distinct results due to the complexity from the yarn. Meridional cotton, even with an inferior quality when compared with the other types, and a lower tenacity, presented a higher elongation, seen in **Figure 3**. It has also been evident that a combined cotton has higher elongation than a carded one.

Regarding rupture force, Meridional cotton had higher values. It is because of a higher count. Other analysis shown a linear behavior, almost uniform, not having force increase or loss, even varying temperature and humidity. As have been proved in the elongation test, combed cotton presented a better performance than the carded cotton.

With regard to rupture work, Meridional cotton carded 660 dtex was analyzed and, as its count is higher than the other types of cotton tested, Meridional shown a rupture work value higher than the others. The other types of cotton presented lower values.

4.2 Count

The count of each sample is obtained through a relation between the fuse weight (grams) and the fuse length (meters). The COATS equipment used provided the weight of the sample (grams) and automatically converted into the count fuse value.

It may be said that there were not great variations on the count medium value. The variations were irrelevant and they may occur because of temperature and humidity changes, whether by the position of the wire on the bobbin, or even by the torsion applied on the spinning mill. There are many reasons for this slight alteration in the count value; however, the temperature and the humidity changes have not altered on a relevant way the bobbins medium value.

TESTES	MERIDIONAL		GIZA 86 CAR		GIZA 86 PENT		GIZA 88	
	CV (%)	σ	CV (%)	σ	CV (%)	σ	CV (%)	σ
TESTE 1	2.704	17.87	2.710	6.089	3.754	8.70	1.049	2.64
TESTE 2	3.057	20.28	1.633	3.656	3.025	7.08	0.915	1.941
TESTE 3	2.252	14.89	2.035	4.634	3.592	8.37	1.041	2.229
TESTE 4	2.047	13.58	1.470	3.329	2.928	6.86	1.090	2.339

Source: Author

Table 4.
Values of count medium variables for each of four cotton types.

Table 4 shows the quotient of count variation and standard deviation. It may be seen through these values that even with this slight modifications, there was no major changes in the medium final result, and therefore it may be concluded that temperature and humidity alterations have not affected considerably the count.

4.3 Torsion

Torsion process was similar to the count process. An arithmetic mean was carried after analyses in each of every six bobbins. To create the torsion pattern, a quality control table from COATS has been used, and it can be seen in **Table 4**.

Torsion analyses were carried through a torsion machine. The edge of the strip value is important to close the torsion because it is the maximum point which can be twisted and gnarled.

The behavior related to the torsion was similar to the count. There were no expressive variations, except for Meridional cotton; however, this is due to a major part of its formation at the spinning mill and the shape from the bobbins where it is wrapped on.

4.4 Nepiness

Nepiness testes conducted on yarns have revealed higher temperature and humidity influence on count and torsion.

For example, being a noble cotton, Giza 88 presented lower values for nepiness in all tests, mainly, when exposed to a lower temperature.

Even being a carded wire, Giza 86 presented an adequate behavior related to nepiness. Similar as happened with Giza 88 carded, Giza 86 has shown less neps on the second and third days. This is due to humidity decreased on the second day and temperature decreased on the third day.

Even passing through the combing process, which removes short fibers paralleling yarns, combed Giza 86 presented a great quantity of neps, and the appearance of 12-mm long fibers could still be noticed. Due to combing process, long fibers remained and short fibers are eliminated.

Being thicker and having a higher count value than the other types of cotton tested, Meridional cotton wires presented a different perspective in other analyses. This yarn showed a higher short fibers quantity and a negligible amount of long fibers. This fact is due to carding process, which do not remove short fibers like the combing process.

5. Conclusion

From the realization of the tests that rendered graphs and tables, some conclusions could be obtained:

a. Yarn Giza 86 presented different results for carded and combed wires. This is because the combed yarns become more resistant to traction than the carded wire, since short fibers are removed.

b. Count and torsion have not shown significant alterations proving that the temperature and the humidity do not influence these characteristics. However, if humidity value was higher than the one used in the test, it could have been assigned another perspective, regarding the count.

c. The results obtained with the Hairiness tester were directly proportional to the data obtained from the tensile strength test, in other words, the lowest is the resistance to traction, the highest is the short fibers quantity. That result can also be compared with the temperature results, so with the temperature increased, it has been and increasing of both parameters.

d. Cotton Giza 88 and Giza 86 combed have the best results for all the analyzed parameters, this is due to both yarn are combed.

e. Cotton Giza 86 combed has shown to best performer to all parameters analyzed, and it has proven the best for the final purpose, that is to say, sewing thread.

f. Meridional cotton, having a higher count value (thicker), has shown that a higher tensile strength to break as well as a superior resistance to the other are necessary. Due to its count value and stronger characteristics, it could not be used or delicate yarns, being used as rustic sewing threads.

Author details

Clara Silvestre de Souza* and José U.L. Mendes
Universidade Federal do Rio Grande do Norte, Brazil

*Address all correspondence to: clarasilvestr@gmail.com

References

[1] de Souza Pereira G. Introdução à Tecnologia Têxtil: Curso Têxtil Em Malharia e Confecção, Módulo 2. Santa Catarina: Apostila-Centro Federal de Educação Tecnológica de Santa Catarina; 2010. p. 101

[2] Geske GA. Transformação da Fibra de Algodão Em Tecido Plano e Felpudo. Sao Paulo: Acurra; 2009. p. 22

[3] Araújo MC. In: Gulbenkian C, editor. Manual de Engenharia têxtil. Vol. 1. Lisboa: Fundação Calouste Gulbenkian; 1984. p. 694

[4] Abidi N, Cabrales L, Hequet E. Thermogravimetric Analysis of Developing Cotton Fibers. Lubbock, TX: Artigo. Fiber and Biopolymer Research Institute, Dept. of Plant and Soil Science, Texas Tech University. Available from: www.elsevier.com/locate/tca; 2009. p. 6 [Accessed: January 2010]

[5] Brionizio JD, Mainier FB. Avaliação de temperatura e umidade em uma câmara climática. Duque de Caxias: Artigo. Instituto Nacional de Metrologia, Normalização e Qualidade Industrial–UFF; 2006. p. 6

[6] Lazzarini JF, Sabino NP. Estudo Comparativo da resistência Do Fio de algodão Obtido Em fiação Piloto e Em fiação Industrial. Campina: Artigo. Instituto Agronômico de São Paulo, Bragant; 1972. 8 p